# 3rd International Symposium on Earth Observation for Arid and Semi-Arid Environments (ISEO 2016)



IOP Conference Series: Earth and Environmental Science
Volume 74

Dushanbe, Tajikistan
19 – 21 September 2016

ISBN: 978-1-5108-4508-4
ISSN: 1755-1307

**Printed from e-media with permission by:**

Curran Associates, Inc.
57 Morehouse Lane
Red Hook, NY 12571

**Some format issues inherent in the e-media version may also appear in this print version.**



Printed by Curran Associates, Inc. (2017)

For permission requests, please contact the Institute of Physics
at the address below.

Institute of Physics
Dirac House, Temple Back
Bristol BS1 6BE UK

Phone: 44 1 17 929 7481
Fax: 44 1 17 920 0979

techtracking@iop.org

**Additional copies of this publication are available from:**

Curran Associates, Inc.
57 Morehouse Lane
Red Hook, NY 12571 USA
Phone: 845-758-0400
Fax: 845-758-2633
Email: curran@proceedings.com
Web: www.proceedings.com

# Table of contents

## Volume 74

**3rd International Symposium on Earth Observation for Arid and Semi-Arid Environments (ISEO2016)**

**19–21 September 2016, Dushanbe, Tajikistan**

**Accepted papers received: 4 July 2017**
**Published online: 21 July 2017**

### Preface



### Papers

IOP Conf. Series: Earth and Environmental Science **74** (2017) 011001 doi:10.1088/1755-1315/74/1/011001

# The 3rd International Symposium on Earth Observation for Arid and Semi-arid Environments (ISEO 2016)

Preface

The 3rd International Symposium on Earth Observation for Arid and Semi-arid Environments (ISEO 2016) was held in Dushanbe, Tajikistan from September 19 to 21, 2016. The theme of this symposium was "Earth Observation for the Great Silk Road".

ISEO 2016 was organized by the Institute of Remote Sensing and Digital Earth, Chinese Academy of Sciences (RADI) and hosted by the Academy of Sciences of the Republic of Tajikistan (ASRT). More than 80 experts and scholars from nine countries and regions had attended the symposium. This volume contains selected papers contributed by the participants.

The purpose of the symposium is to further exchange experiences on the use of Earth observation technologies among the Central Asian and neighboring countries. In 2012, the 1st symposium in this series was initiated in Kashgar by RADI, CAS. The 2nd symposium was held in 2014 in Issyk Kul, Kyrgyzstan. Since then, an international network has been established preliminarily to promote cooperation on Earth observation for arid and semi-arid environment, especially in Central Asia and neighboring regions.

ISEO 2016 had provided a scientific platform to discuss and explore better scientific approaches to global sustainable development in the Belt and Road region through Earth observation technologies. I do hope that we could have more bilateral and multi-lateral scientific cooperations among the Belt and Road countries with the ISEO symposia.

GUO Huadong
Co-Chair, ISEO 2016
Academician, Chinese Academy of Sciences (CAS)
Institute of Remote Sensing and Digital Earth (RADI), CAS

# The symposium mainly include the following topics:

Earth Observation: Theory and Technology
Earth Observation for Eco-environment and Urbanization
Earth Observation for Heritage
Earth Observation for Disaster Mitigation
Earth Observation and Climate Change
Earth Observation for Mineral, Energy and Water Resources

# Editors of the ISEO2016 Proceedings

Prof. Bihong Fu
Prof. Li Jia
Prof. Jimin Sun
Prof. Xinwu Li
Prof. Pilong Shi

# List of Committee Members

# COMMITTEES

| | | |
|---|---|---|
| Symposium Co-Chairs: | Farhod Rahimi | Academician, President of the Academy of Sciences of the Republic of Tajikistan |
| | Huadong GUO | Academician, Institute of Remote Sensing and Digital Earth, Chinese Academy of Sciences |

# INTERNATIONAL SCIENTIFIC COMMITTEE

| | | |
|---|---|---|
| Chairs: | Abdusattor Saidov | Chief Scientific Secretary, Academy of Sciences of the Republic of Tajikistan |

| | | |
|---|---|---|
| Members: | Muhiba Yakubova | Academician, Vice-President, Academy of Sciences of the Republic of Tajikistan |
| | Kobuliev Zaynolobuddin | Director, Institute of Water problems, Hydro-energy and Ecology, Academy of Sciences of the Republic of Tajikistan |
| | Khikmat Muminov | Director, Physics and Technical Institute, Academy of Sciences of the Republic of Tajikistan |
| | Akobir Mirzorahimov | Director, Institute of Botany, Plant Physiology and Genetics, Academy of Sciences of the Republic of Tajikistan |
| | Faizali Khakimov | Director, Institute of Zoology and Parasitology, Academy of Sciences of the Republic of Tajikistan |
| | Kubanychbek Zhumaliev | Academician, Kyrgyz National Academy of Sciences |
| | Sakiev Kadyrbek | Director, Institute of Geology, Kyrgyz National Academy of Sciences |
| | Abdrakhmatov Kanatbek | Director, Institute of Seismology, Kyrgyz National Academy of Sciences |
| | Marat Omurserikov | Director, K.I. Satpaev Institute of Geological Sciences, Kazakhstan |
| | Tanatkan Abakanov | Director, Institute of Seismology, Kazakhstan |
| | Abdulla Saparov | Director, U.U.Uspanov Kazakh Research Institute of Soil Science and |

| | | |
|---|---|---|
| | | Agrochemistry, Kazakhstan |
| | Sabriddin Khisomidinov | Director, Institute of Seismology of Academy of Sciences, Uzbekistan |
| | Mirzaev Abdurazak | Director, Institute of Geology and Geophysics of Academy of Sciences, Uzbekistan |
| | Hans-Bandel Havenith | University of Liege, Belgium |
| | Tandong YAO | Academician, Institute of Tibetan Plateau Research, CAS |
| | Congqiang LIU | Academician, National Natural Science Foundation of China |
| | Jimin SUN | Institute of Geology and Geophysics, CAS |
| | Bing ZHANG | Institute of Remote Sensing and Digital Earth, CAS |
| | Gensuo JIA | Institute of Atmospheric Physics, CAS |
| | Jianbo LIU | Institute of Remote Sensing and Digital Earth, CAS |
| | Xi CHEN | Director, Xinjiang Institute of Ecology and Geography, CAS |
| | Rasul Ghulam | Director, Pakistan Metrology Bureau |

# ORGANIZING COMMITTEE

| | | |
|---|---|---|
| Co-Chairs: | Yunus Mamadjanov | Institute of Geology, Earthquake Engineering and Seismology, Academy of Sciences of the Republic |

| | | |
|---|---|---|
| | | of Tajikistan |
| | Bihong FU | Professor, Director of Kashgar Research Center, RADI, CAS |
| Secretary-General: | Ilhomjon Oimahmadov | Director, Institute of Geology, Earthquake Engineering and Seismology, TAS |
| Members: | Natarajan Ishwaran | Institute of Remote Sensing and Digital Earth, CAS (Sri lanka) |
| | Changlin WANG | International Society for Digital Earth |
| | Tianhua HONG | International Centre on Natural and Cultural Heritage using Space Technology under the auspices of UNESCO |
| | Fang CHEN | CAS-TWAS Centre of Excellence on Space Technology for Disaster Mitigation |
| | Farkhod Hakimov | Institute of Geology, Earthquake Engineering and Seismology, Academy of Sciences of the Republic of Tajikistan |
| | Shohrukh Murodkulov | Institute of Geology, Earthquake Engineering and Seismology, Academy of Sciences of the Republic of Tajikistan |
| | Xinwu LI | Institute of Remote Sensing and Digital Earth, CAS |
| | Jie LIU | Institute of Remote Sensing and Digital Earth, CAS |
| | Yoshiki Ninomiya | Geological Survey of Japan, AIST, Japan |

IOP Conf. Series: Earth and Environmental Science **74** (2017) 011001 doi:10.1088/1755-1315/74/1/011001

| | | |
|---|---|---|
| | Yunpeng WANG | Guangzhou Institute of Geochemistry, CAS |
| Secretariat: | Jovid Aminov | Institute of Geology, Earthquake Engineering and Seismology, Academy of Sciences of the Republic of Tajikistan |
| | Madina Dzhursizova | Institute of Geology, Earthquake Engineering and Seismology, Academy of Sciences of the Republic of Tajikistan |
| | Mingmei CHEN | Institute of Remote Sensing and Digital Earth, CAS |
| | Pilong SHI | Institute of Remote Sensing and Digital Earth, CAS |
| | Jing ZHANG | Institute of Remote Sensing and Digital Earth, CAS |

IOP Conf. Series: Earth and Environmental Science **74** (2017) 011001 doi:10.1088/1755-1315/74/1/011001

IOP Conf. Series: Earth and Environmental Science **74** (2017) 011001
doi:10.1088/1755-1315/74/1/011001

# Peer review statement

All papers published in this volume of *IOP Conference Series: Earth and Environmental Science* have been peer reviewed through processes administered by the proceedings Editors. Reviews were conducted by expert referees to the professional and scientific standards expected of a proceedings journal published by IOP Publishing.

IOP Conf. Series: Earth and Environmental Science **74** (2017) 012001 doi:10.1088/1755-1315/74/1/012001

# Some characteristics of ozone concentrations and their linkages with meteorological factors in Kashgar, northwestern China

**Huiqin Wang, Mamtali Mamtimin, Feng Chen,**
Key Laboratory of Tree-ring Physical and Chemical Research of China Meteorological Administration, Institute of Desert Meteorology, China Meteorological Administration, Urumqi 830002, China

E-mail: huiqinwang1@163.com

**Abstract.** Observations of $O_3$ and $NO_2$ have been performed in Kashgar, northwestern China, from January to December 2015. We analyze the variations of $O_3$ and $NO_2$ daily, the diurnal variations of $O_3$ and $NO_2$ as well as the relation between $O_3$ and $NO_2$ concentrations. We focused on the variations of $O_3$, with the meteorological factors. We find that there exist to some extent between meteorological factors and $O_3$ concentrations. High $O_3$ concentrations were related to high temperature, low relative humidity, low cloudiness, and light wind speed. The local meteorological conditions on certain situations also influenced the ozone concentration. Preliminary statistics of the occurrence frequencies of high ozone concentration days corresponding to some meteorological factors are given in this paper.

## 1. Introduction

The conditions and sources for producing photochemical smog have been discussed extensively recently [1,2]. As the most important index of photochemical smog, ozone has been thought as one of the primary pollutants influencing air quality. Kashgar is one of the largest cities in Central Asia, where a large quantity of nitrogen oxides, precursor and hydrocarbons are emitted. With the population growth and socio-economic development in Kashgar, the threat of air pollution due to photochemical smog is growing. Thus, we have made a series of simultaneous observations of $O_3$ and $NO_2$ in the lower atmosphere in Kashgar since 2015. In this paper, we analyze the characteristics of $O_3$ concentrations and compare it with meteorological factors in order to find some conditions favoring production of high ozone concentration. Ozone concentration measurement system (ML9810, Monitor Labs Company, Australia) and Nitrogen Oxides Analyser (ML9841, Monitor Labs Company, Australia) was used to the measurements of $O_3$ and $NO_2$.

## 2. Relations between $NO_2$ and $O_3$ concentration

In 2015, we performed 345 d of observations of $O_3$ and $NO_2$ simultaneously at a height of 3 m above the ground. The measurement site was located at the southern district of Kashgar City. Figure 1 shows the concentration curves of $O_3$ and $NO_2$ that the $O_3$ curve basically corresponded to the $NO_2$ curve. The correlation coefficients between the individual concentrations of $O_3$ and $NO_2$ were -0.64 for the levels of 3 m. This implies that the $O_3$ concentrations will be higher when the $O_3$ precursors concentration declined, such as $NO_2$, which is very similar to that derived from the correlation between $O_3$ and $NO_2$ in eastern China [3]. Figure 1 also shows the curves of $O_3$ and $NO_2$ were less consistent with each other in winter, especially in February.

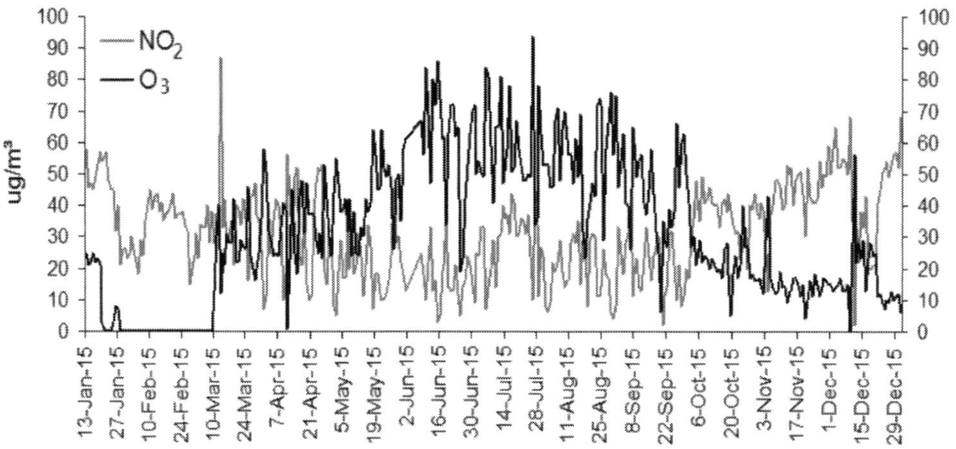


Figure.1 Daily mean $O_3$ and $NO_2$ concentrations during measurement period at Kashgar station.

## 3. Relationships between meteorological factors and daily concentrations of $O_3$

The daily mean concentration of ozone fluctuated significantly. On July 26th, for example, the highest values of mean concentrations were 94 μg/m$^3$. On July 27th, a clear day on which light fog appeared in the morning. The $O_3$ concentration at the Kashgar observatory was also characterized by the obvious seasonal variation. The $O_3$ concentrations in spring (Mar, Apr and May), summer (Jun, Jul and Aug), autumn (Sep, Oct and Nov), and winter (Dec, Jan and Feb) were 31.8, 58.1, 27.3, and 10.5μg/m$^3$, respectively, with the order of winter＜spring＜autumn＜summer. We consider that the strong variations of the ozone concentrations were mainly caused by meteorological conditions. Ozone concentrations produced from photochemical reactions were influenced by meteorological factors (Table 1; significant correlations ($p$<0.001) indicated in bold font). Ozone concentrations were positively correlated with temperature, suggesting that high temperature led to an increase in the formation of $O_3$. The favorable meteorological conditions (mild winds, high temperature, and clear skies) have been reported to exert a great influence on $O_3$ levels (Vecchi and Valli 1999). In the present study, a significant negative correlation coefficient was found between $O_3$ concentration and relative humidity. These results indicate that high relative humidity with strong solar radiation were important factors causing higher $O_3$ concentrations during the observation period.

Table 1 Correlation coefficients between $O_3$ concentration, temperature, relative humidity, solar radiation, mean wind speed and cloudiness during the observation period

| | N | Temperature | Relative humidity | Solar radiation | Mean wind speed | Cloudiness |
|---|---|---|---|---|---|---|
| $O_3$ | 345 | 0.85* | -0.42* | 0.51* | 0.33* | 0.27* |

* Significant at the 1% level.

## 4. Daily variations of $NO_2$ and $O_3$ concentration

To analyze the feature of diurnal variations, the data of observations made twelve times in daytime and twenty-four times through a whole day are used. The results show that the peaks of $NO_2$, appeared at 10-12h and the peaks of $O_3$, occurred at 14-16 h. The occurrence of the $O_3$, peak value was 4h later than that of $NO_2$. Generally, a peak value of $O_3$, occurred only once during the daytime using the data of twelve times a day; however, 24 h measurement data show that a secondary peak value of $O_3$, existed at 0-2 h as shown in Figure 2. The meteorological parameter curve correspond clearly to the ozone concentration curve (Figure 2). Temperature usually reached a maximum value around 14-15 h while the $O_3$ peak value occurred immediately after midday. This might be related to the fact that vertical transfer became active with the development of mixing, perhaps bringing ozone from upper to lower levels [5]. This would cause abnormal increase of $O_3$ concentration in the atmosphere.

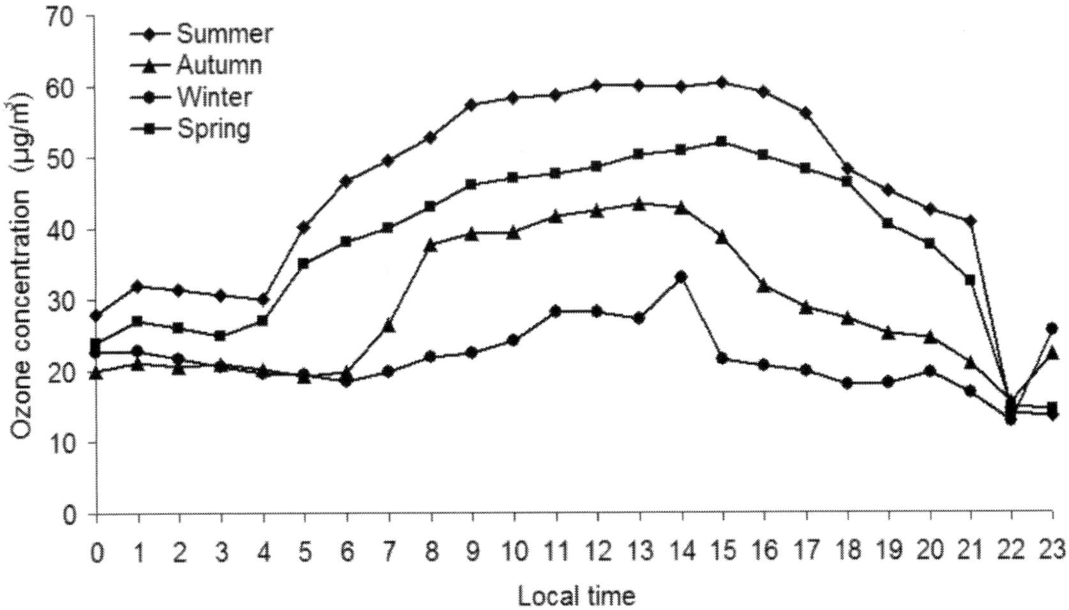


Figure.2 Variation of the hourly mean $O_3$ concentration in four seasons

## 5. Summary

The variations of $O_3$ concentration were basically linkages with the $NO_2$ concentration. Correlations existed to some extent, between meteorological factors and $O_3$ concentrations. High concentrations of $O_3$ were related to high temperature, light wind, low relative humidity and low cloudiness. In daytime, the occurrence of the $O_3$ peak was later by about 4 h than that of the $NO_2$ peak. At night, a secondary peak of $O_3$ existed at 0-2 h. High daily concentrations of ozone usually occurred in the situation of high temperature when the local photochemical reaction mainly caused the increase of ozone in the lower atmosphere.

**Acknowledgments**
This study was financially supported by Central Asian Atmospheric Science Research Fund (CASS201714).

## References

[1] Aidaoui L, Triantafyllou A G, Azzi A, Garas S K and Matthaios V N 2015 Elevated stacks' pollutants' dispersion and its contributions to photochemical smog formation in a heavily industrialized area *Air Qual., Atmos. Health* **8** 213-27
[2] Baier B C, Brune W H, Lefer B L, Miller D O and Martins D K 2015 Direct ozone production rate measurements and their use in assessing ozone source and receptor regions for Houston in 2013 *Atmos. Environ* **114** 83-91
[3] Xie Y, Zhao B, Zhang L and Luo R 2015 Spatiotemporal variations of $PM_{2.5}$ and $PM_{10}$ concentrations between 31 Chinese cities and their relationships with $SO_2$, $NO_2$, CO and $O_3$ *Particuology* **20** 141-49
[4] Vecchi R, and Valli G 1999 Ozone assessment in the southern part of the Alps *Atmos. Environ.* **33** 97-109
[5] Xu X, Lin W, Wang T, Yan P, Tang J, Meng Z, and Wang Y 2008 Long-term trend of surface ozone at a regional background station in eastern China 1991–2006: enhanced variability *Atmos. Chem. Phys.* **8** 2595-607

IOP Conf. Series: Earth and Environmental Science **74** (2017) 012002 doi:10.1088/1755-1315/74/1/012002

# Glacier Changes in the West Kunlun Mountains Revealed by Landsat Data from 1994 to 2016

**J Zhang[1], B H Fu[1*], L M Wang[1], A Maimaiti[1], Y X Ma[1] and F Yan[1]**
[1]Institute of Remote Sensing and Digital Earth (RADI), Kashgar Research Centre, Chinese Academy of Sciences (CAS), Beijing 100094, P. R. China

E-mail: fubh@radi.ac.cn

**Abstract:** The dynamics of glacier area in the West Kunlun Mountains has been measured using Landsat TM, ETM+, and OLI data form 1994 to 2016. The study mainly focuses on the northern slope of West Kunlun Mountains, a region that is affected by the Westerlies. From 1994 to 2016, glacier area slightly increased then decreased dramatically in the whole period. The most significant glacier shrinkage occurred in 2005. Before 2005, the glacier was growing, with the area increased from 1687.44 $km^2$ in 1994 to 1846.67 $km^2$ in 2005. After 2005, the glacier retreated rapidly, with a recession rate of 21.31 $km^2 a^{-1}$ during 2005-2008. From 2008 to 2016, the glacier continued to shrink with a recession rate of 11.00 $km^2 a^{-1}$, and the glacier area finally declined to 1694.57 $km^2$ in 2016. Finally, combining the classification result with DEM, we estimated that the average transient snowline altitude was 6000 m in the study area.

## 1. Introduction

Mountain glaciers are a vital water source for arid and semi-arid regions, their variations are crucial to the livelihood of the downstream areas. More than 6.3 million people rely on the fresh melting water from West Kunlun Mountain glaciers for drinking, agriculture, and hydropower [1]. Meanwhile, mountain glaciers are commonly regarded as a sensitive indicator of local or global climate change which is closely related with the hydrological cycle, snow accumulation, evapotranspiration, runoff and recharge to the subsurface [2]. Knowledge of the dynamics of glacier area will help us alleviate problems such as glacier lake outbursts, floods, water shortage, and environmental vulnerability.

Remote sensing is a useful tool for monitoring changes in glacier boundaries on multi-decadal time scales. Compared to the systematic field-based measurements, remote sensing is cheaper, easier, and safer. Although almost all the remote sensing techniques require some kind of ground control or validation, this approach provides possibility for larger spatial coverage, which is important for temporal-spatial monitoring of land surface. Actually, remote sensing is the only practical approach of obtaining statistically meaningful sample of glacier mass balance (m.b.) estimates at a regional scale [3].

The surface of glacier shows different optical properties because it is composed of snow, firn, ice, water and debris (e.g. rock, pebbles, dust, soot) [4]. Accumulation and ablation zones can be detected through the characteristic features in the surface layers of a glacier. Based on their unique optical properties, ice cover from glaciers can be separated from snow, firn and moraine/bedrock. Furthermore, the equilibrium line altitude can be obtained by the aid of digital elevation model (DEM). The specific objectives of this study are to a) map and classify the glacier zones on the northern slope of the West Kunlun Mountains; and b) derive

* Corresponding author.

the snow line altitude as the proxy for the altitude of equilibrium line according to previously mapped distribution of glacier zones.

## 2.Methods

### *2.1 Study Area*

This study was focused in the West Kunlun Mountains (35°-36°N, 80°-82°E), southern Xinjiang, adjacent to the Tarim Basin in the north. The closed Tarim Basin have two main rivers: Yurungkax (Yurung) River and Keriya River. The West Kunlun glacier supplies most of annual runoff of these two rivers [5], therefore this study are of great significance. The climate of this region is dominated by Westerlies, and the local glaciers are regarded as continental-type [6]. The elevation of the glaciers ranges from 5000 to 7000 m above sea level (a.s.l.). Since 2000, the average annual temperature and annual precipitation have been −10 °C and ~ 500 mm, respectively [7] above 5800 m a.s.l. Many glaciers developed due to the cold climate in this region.

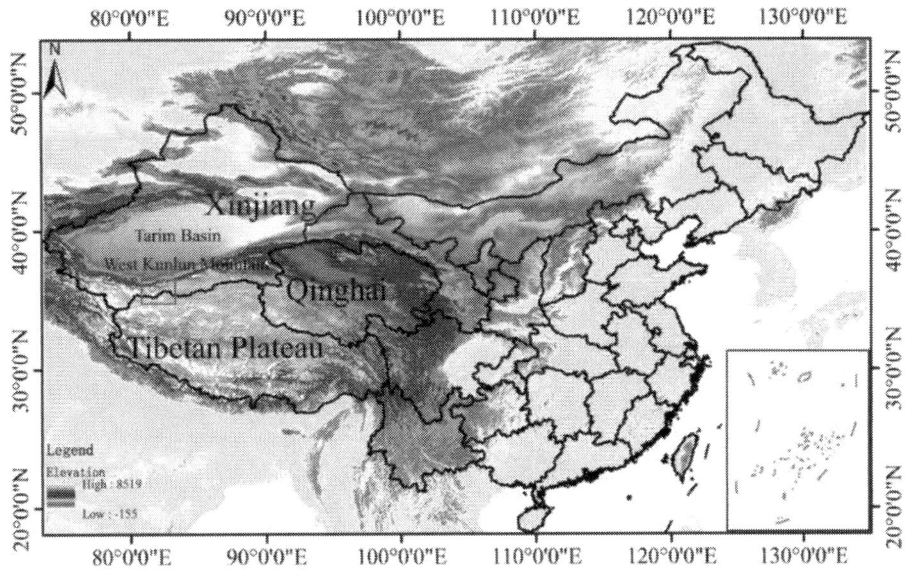


**Figure 1.** Location of the study area in the West Kunlun Mountains.

### *2.2 Data*

Orthorectified and projected Landsat TM/ETM+ surface reflectance data with spatial resolution of 30m×30m were downloaded from the United States Geological Survey (USGS) Earth Resources Observation and Science (EROS) Center. Additionally, Landsat OLI L1T (level-1 terrain-corrected) data were also used in the study. For detailed study of glacier area variation, we selected scenes (path/row: 145/35) from the following years: 1994 (Landsat 5 TM), 2005 (Landsat 7 ETM+), 2008 (Landsat 7 ETM+) and 2016 (Landsat 8 OLI). All scenes were acquired in June to ensure the minimum of fresh snow cover. Each scene was under the condition of minimal cloud cover and haze. All Landsat data were projected to the Universal Transverse Mercator (UTM) coordinate system, zone 44 north. The original Landsat 8 OLI L1T data was converted to surface reflectance using the FLAASH (the Fast Line-of-sight Atmospheric Analysis of Spectral Hypercubes) atmospheric correction module embedded in the ENVI (Environment for Visualizing Images, Version 5.1) software.

Shuttle Radar Topography Mission (SRTM) 90m resolution Digital Elevation Data (DEM) was downloaded from the Consultative Group for International Agriculture Research — Consortium for Spatial Information (CGIAR; http://srtm.csi.cgiar.org/). The DEM was used for generating contour lines, dividing watershed, and analyzing slope of glaciers.

### *2.3 Decision Tree classification*

The Decision Tree classification uses a hierarchical splitting mechanism and has been widely used in remote sensing. In this study, in order to gain a comprehensive understanding of debris-free and debris-covered glaciers, we applied this method on multispectral image data. Based on knowledge of the spectral properties of land surface, we established a tree structure to identify the debris-free and debris-covered glacier areas. Using the reflectance value as input, we categorize each pixel into debris-free glacier (clean ice), debris-covered (dirty ice/snow) bare land and shadow.

For establishing a decision tree, a training set of ROI (Region of Interest) are firstly chosen from Landsat scenes which are tagged with decision labels and contain a group of attribute values. The C5.0 model used 70% of the input datasets to determine the optimal thresholds of each indicator, and then to separate pixels into different classes [4]. The left 30% of the input sample datasets were used for evaluating the classification accuracy. Only under certain minimum errors, the tree-generating process will give the optimal thresholds, otherwise the process will be repeated. Further details about this method can be found elsewhere [8,9]. Finally, the obtained hierarchically structured rules were applied to classify the glacier surfaces of different periods.

## 3. Results and Discussion

### *3.1 Analysis of classification results*

Results of the decision tree classification are summarized in Tables 1 and 2. For each year, the hierarchical clustering algorithm computed five classes of glacier surface types: debris-free glacier, debris covered glacier, bare land type 1, bare land type 2, and shadow. Fig. 2 shows the classification results for all four periods. In this study, the debris-free and debris covered glaciers were the two main objects being investigated. Therefore, the two kinds of bare land soils were set in the same color in Fig.2.

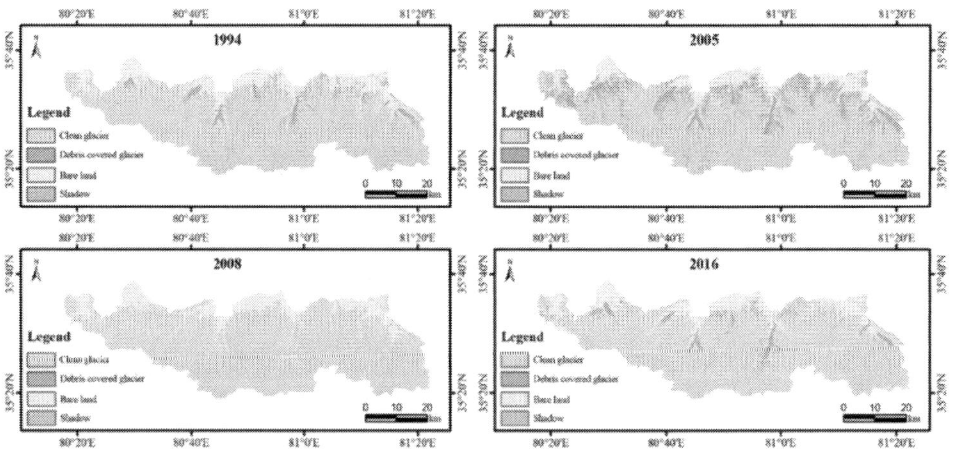


Figure 2. Classified glaciers of the West Kunlun Mountains form 1995 to 2016.

**Table 1.** Changes in debris-free, debris-covered, and the total ice areas of the West Kunlun Mountain regions from 1994 to 2016, respectively.

| Year | Debris-free ice | | Debris-covered ice | | Total area (km²) |
|---|---|---|---|---|---|
| | Area (km²) | Percentage (%) | Area (km²) | Percentage (%) | |
| 1994 | 1608.35 | 95.3 | 79.10 | 4.7 | 1687.44 |
| 2005 | 1542.53 | 83.5 | 303.94 | 16.5 | 1846.47 |
| 2008 | 1779.64 | 99.84 | 2.90 | 0.16 | 1782.54 |
| 2016 | 1611.20 | 95.08 | 83.36 | 4.92 | 1694.57 |

**Table 2.** The dynamics of total ice area of the West Kunlun Mountain regions in four periods.

| Period | Area change (km²) | Mean annual rate ($km^2 a^{-1}$) | Area change (%) |
|---|---|---|---|
| 1994-2005 | 159.03 | 14.46 | 9.42 |
| 2005-2008 | -63.93 | -21.31 | -3.46 |
| 2008-2016 | -87.98 | -11.00 | -4.94 |
| 1994-2016 | 7.12 | 0.32 | 0.42 |

As shown in Tables 1 and 2, the glacier area from 1994 to 2016 displayed a slightly increase of 0.42%, from 1687.44 $km^2$ in 1994 to 1694.57 $km^2$ in 2016. The glacier area expanded at a rate of 14.46 $km^2$ $a^{-1}$ between 1994 and 2005. Conversely, the glacier rapidly retreated in the northern slope of the West Kunlun Mountains from 2005 to 2008 with an annual rate of 21.31 $km^2$ $a^{-1}$ (Table 1). This receding trend continued in a rate of 11.00 $km^2$ $a^{-1}$ during the period 2008–2016.

*3.2 The snow lines*

According to the classification result, most of the dark surfaces distribute at the middle to downwards regions. The transient snow line is easily visible on the classified images. Basically, the surface reflectance separates the darker bare ice from the bright snow. Then, we generated contour lines with 20m intervals based on SRTM data. Combing the contour line with the glacier classification zone, the average transient snow line can be define as 6000 m. Zhang and Jiao (1987) reported that the snow line of southern slope of the West Kunlun Mountains was 5900 m [10]. Nakawo (1990) published the elevation of the Chongce Ice Cap (in West Kunlun) terminus is about 5800 m [11]. The snow line tends to the higher elevation in recent years.

## 4. Conclusion

This study aims to investigate the temporal behavior of glacier surfaces using remote sensing techniques to classify the different zones of glaciers. Using multi-temporal Landsat images as input data, the decision tree classification method was employed to classify glacier surfaces and to monitor their change. Following conclusions can be drawn:

1) There was no significant change in glacier area in the study area as a whole, and the glacier area slight increase about 7.12 $km^2$ (0.42%) over the 22 years. But the glacier changes were not homogeneous.

2) In the context of global warming effect, the average snow line is becoming higher. Further studies are still needed to better understand the elevation of the snow lines in areas like the West Kunlun Mountains.

**Acknowledgements**

This research is financially supported by Chinese Academy of Sciences "Light of West China" Program （2015-XBQN-A-07）

**References**

[1] Li C X, Yang T B, Tian H Z 2015 Variation of Western Kunlun Mountain glaciers monitored by remote sensing during 1976-2010 *Mountain Research* **2** 157-65 (In Chinese with English abstract)

[2] IPCC (International Panel for Climate Change): http://www.ipcc.ch/report/ar5/syr.

[3] Bamber J L and Rivera A 2007 A review of remote sensing methods for glacier mass balance determination *Global Planet Change* **59** 138-48

[4] Pellikka P K E and Rees G 2010 Remote sensing of glaciers: techniques for topographic, spatial, and thematic mapping of glaciers *J. Glaciol.* **56** 927

[5] Bao W J, Liu S Y, Wei J F, et al. 2015 Glacier Changes during the Past 40 Years in the West Kunlun Shan *J. Mt. Sci.* **12** 344-57

[6] Shi Y F 2008 *Concise Glacier Inventory of China* (Shanghai: Shanghai Popular Science Press)

[7] Wu H. Regional glacier mass loss estimated by ICESat-GLAS data and SRTM digital elevation model in the West Kunlun Mountains, Tibetan Plateau, 2003–2009 2014 *J. Appl. Remote Sens.* **8** 083515

[8] Quinlan J R 1999 Simplifying decision trees *Int. J. Hum-Comput. St.* **51** 497-510

[9] Mingers J 1989 An empirical comparison of pruning methods for decision tree induction *Mach. Learn.* **4** 227-43

[10] Zhang Z and Jiao K 1987 Modern glaciers on the south slope of West Kunlun Mountains (in Akasyqin Lake and Guozha Co Lake drainage areas) *Bull. Glacier Res.* **5** 85–91

[11] Nakawo M 1990 Climatic information from the Chongce Ice Cap, West Kunlun, China *Ann. Glaciol.* **14** 205-07

IOP Conf. Series: Earth and Environmental Science **74** (2017) 012003 doi:10.1088/1755-1315/74/1/012003

# Development of urbanization in arid and semi arid regions based on the water resource carrying capacity -- a case study of Changji, Xinjiang

**Xiao H[1], Zhang L[1], Chai Z[2]**
[1] Institute of Remote Sensing and Digital Earth, Chinese Academy of Sciences, No.9 Dengzhuang South Road, Haidian District, Beijing 100094, China
[2] North China University of Science and Technology,46 Xinhua Road, Tangshan 063009, Hebei, China

E-mail:xiaohan@radi.ac.cn

**Abstract.** The arid and semiarid region in China where have a relatively weak economic foundation, independent development capacity, and the low-level of urbanization. The new urbanization within these regions is facing severe challenges brought by the constraints of resources. In this paper, we selected the Changji Hui Autonomous Prefecture, Xinjiang Uyghur Autonomous Region as study area. We found that agricultural planting structure is the key water consumption index based on the research about the main water demands of domestic, agriculture and industry. Finally, we suggest that more attentions should be paid to the rational utilization of water resources, population carrying capacity, and adjust and upgrade the industrial structure, with the purpose of coordination with the Silk Road Economic Belt.

## 1. Introduction

In 1935, geographers Hu Huanyong proposed a China's population density contrast line [1]. The line divided China into two parts. The area of the southeast of the line with population of more than 96% of the country in an area of 36% of the total area. The area of the northwest of the line with population of 4% and an area of 64% of the country [2]. In the meantime, this line fits with China's 400 mm annual rainfall isoline and is boundary of China semi-arid region and semi-humid region [3]. The climatic and environmental constraints directly led to the fact that a vast territory with a sparse population and weak economics in the northwest arid and semi-arid region [4].

Rapid urbanization and industrialization would inevitably cause pressure on the regional ecological environment [5]. Only quickly upgrading the level of urbanization and industrialization effectively alleviating the pressure on the fragile ecological environment and water resources in the western region of China, can ensure the sustainable development of urbanization and industrialization[6]. Chinese president Xi Jinping proposed the Grand strategic idea about the Silk Road Economic Belt offer the new opportunity for the development of the arid and semi-arid regions in western China [7]. We selected the Changji Hui Autonomous Prefecture, Xinjiang Uyghur Autonomous Region as study area, under the condition of limited water resources, to put forward a more reasonable new urbanization structure framework.

---

[1] Han XIAO, research assistant, Institute of Remote Sensing and Digital Earth, Chinese Academy of Sciences.

## 2. Data & Methodology

### *2.1. Data*

This research used Landsat 8 OLI_TRIS satellite digital products with spatial resolution of 30 meters and DEM with spatial resolution of 30 meters. They were both from Geo spatial data cloud (http://www.gscloud.cn/). Data pre-processing include image stitching and cutting, atmospheric correction, etc. Demographic and economic data were from Bureau of Statistics of Changji Hui Autonomous Prefecture. Water supply and use data were from the Statistics Bureau of Xinjiang Uygur Autonomous Region. The cultivated area and the unit area yield data of the main crops were from the Changji Hui Autonomous Prefecture national economic and social development statistical bulletin in 2015.

### *2.2. Methodology*

#### *2.2.1. Statistical Study of Changji's urbanization*

In recent years, Changji's economic and social developed rapidly. In 2015, the GDP of Changji was 114 billion yuan and increased by 12% over the previous year. Among them, the first industry added to 25 billion 200 million yuan, increased by 4.9%; second industry added value of 55 billion yuan, increased by 15.3%; the third industry added value of 33 billion 800 million yuan, increased by 11.2%. Per capita GDP of the resident population was 70700 yuan and increased by 12.0% [8]. But economic was imbalance at Changji where western two counties and one city accounted for 68% of the total economic output, and eastern three counties accounted for 32%.

#### *2.2.2. Statistical Study of Water resources in Changji*

The total amount of available water resources is 2 billion 942 million cubic meters in the Changji Hui Autonomous Prefecture, accounting for only 4.04% of the available total amount of the whole Xinjiang. Changji is one of the most water shortage states in Xinjiang. The per capita possession of surface water resources is only 1,617 cubic meters and is below the average level of the whole country (1920 cubic meters) and Xinjiang (2988 cubic meters) and is below the Internationally recognized signs point for water use (1700 cubic meters). This is visible that water resources pressure is obvious in the development process of Changji[9],[10].

**Table 1.** Water resources comparison between Chingji and Urumqi（Units: billion cubic meters）.

| Region | Water supply | Water consumption | | | |
|---|---|---|---|---|---|
| | | primary industry | secondary industry | tertiary industry | Life & environment |
| Urumqi | 11.24 | 6.51 | 2.39 | 0.32 | 2.01 |
| Hui Autonomous Prefecture of Changji | **47.92** | **44.27** | **2.20** | **0.16** | **1.30** |

(Data sources: Statistical Bureau of Urumqi Autonomous Region)

The primary industry precise agricultural irrigation water accounts for 91% of the total water consumption of the national economy. It plays a significant role in the water structure of the Hui Autonomous Prefecture of Changji. Israel has a more severe problem about water resource shortages, but Israel's agriculture has become one of the three pillar industries for the national economy and foreign trade[11-13].

**Table 2.** The agricultural economic contrast analysis between Changji and Israel

| Region | Crop | Cultivated Area($hm^2$) | Water consumption per unit($m^3/hm^2$) | Yield(kg/ $hm^2$) | Water economic benefit (USD/$m^3$) |
|---|---|---|---|---|---|
| Hui Autonomous Prefecture of Changji (in 2015) | wheat | 185720 | 5 742 | 5936.85 | 0.41 |
| | corn | 108240 | 5052 | 10296.6 | 0.52 |
| | cotton | 80286.67 | 8 650 | 1867.05 | 0.27 |
| | primary industry | **547366.7** | \ | \ | **1.22** |
| Israel (in 1997) | \ | **518100** | **2179** | \ | **13.59** |

(Data sources: Source Mews Beauru of Israel Embassy in China, "Agriculture of Israel (1997)" )

*2.2.3. Constructed water resources carrying capacity model*

Although the total amount of water resources in the Hui Autonomous Prefecture of Changji is relatively stable, the water supply capacity in Changji region continued to increase from 2005 to 2014 years, especially after the year of 2010. In 2014, the total water consumption in Changji was 4 billion 792 million cubic meter, which is about 4.3 times of Urumqi City, while Changji's GDP was 106 billion yuan, which was less than 50% of Urumqi's GDP. The first industrial water accounted for 92.4% of the total water supply, second and third of the industrial water accounted for 4.9% of the total water supply, water consumption accounted for only 1.7% of the residents. Among them, the agricultural irrigation water accounted for 91% of the total water consumption of the national economy. Agricultural water consumption has a significant role on the total amount of water in Changji, and is also the key factor to restrict the carrying capacity of water resources in the Hui Autonomous Prefecture of Changji. Although the water structure has improved slightly in recent years, the change is not significant.

The structure of water consumption can reflect the economic level, the scientific and technological levels, and the civilization degree. In developed countries, the proportion of living water is 15%~20%, the proportion of industrial water is 35%~40%, and the proportion of agricultural water is 45% ~ 50%[14].

We use regression analysis method and nearly a decade of data sets built the per capita water demand forecasting model of the Hui autonomous prefecture of Changji, then improve the model by selecting some coefficients from the structure of water consumption in developed country, the final conceivable prediction population capacity based on water resources may attain to 2.0514 million at Changji. This number may be a good reference in the future development of urbanization in this region.

## 3. Conclusion

Changji's situation of water supply has improved in recent years, water supply capacity increased year by year. But the supply of water resources is basically used by the first industry which did few contributions to the region economic development. Adjusting the water supply structure, reducing the first industrial water consumption, more to use on the second industries, promoting the transformation of economic structure, is conducive to better development.

**Acknowledgments**
This work was supported by National Natural Science Foundation under Grant 41501475; and Project of Institute of Remote Sensing and Digital Earth.

**References**

[1] Hu H Y 1990 The distribution, regionalization and prospect of China's population *Acta Geogr. Sin.* **45** 139–14 (In Chinese with English abstract)
[2] Zhang F 2014 *China's Urbanization and the World Economy* (Aldersbot: Edward Elgar

Publishing) pp 347-367

[3] Wang Z, Zhang P Y, Zhou Q B 1996 The influence of historical climate change on social development in China-on the relationship between man and land *Journal of Geography* **51** 329-39

[4] Qian J X 1990 *The introduction of Master Hu Huanyong. In: Wu Chuanjun,Shi Yafeng.A Memoir of the Development of Chinese Geography for the Past 90 Years* (Beijing: Xueyuan Press) pp 50-55

[5] Jun W 2003 Land Degradation and Ecological Rehabilitation in Karst Areas of Guizhou Province,South Western China *Advances in Earth Science* **18** 447-53 (In Chinese with English abstract)

[6] Liu J, Li S et al. 2008 Ecological and Socio-economic Effects of China's Policies for Ecosystem Services *PNAS* **105** 9477–9482

[7] National Development and Reform Commission, Ministry of foreign affairs, Ministry of Commerce 2015 The Belt and Road vision and action

[8] Statistical Bureau of Changji Hui Autonomous Region 2015 Analysis on the operation of national economy in Changji in 2015

[9] Cui L L 2008 Status quo and existing problems of groundwater exploitation and utilization in Changji Prefecture, Xinjiang *Heilongjiang water conservancy science and technology* **02** 123-24 (In Chinese with English abstract)

[10] Liu Z H,Wang H J,Pei H,Tian W R, Ju C Y,Zhou X 2005 Analysis of glacier changes in the state of Changji, Xinjiang, based on RS and GIS Technology *Journal of Xinjiang University* **02** 127-33

[11] Fishelson C 1994 The allocation of the Marginal Value Product of water Israel Agriculture. *Water and Peace in the Middle East* 427-440

[12] Chapagain A K Hung P Q 2003 Virtual Water Trade: a quantification of virtual water flows between nations in relation to international crop trade *J.org.chem* **11** 835–55

[13] Chapagain A K Hoekstra A Y 2003 *Virtual water Trade: a quantification of virtual water flows between nations in relation to international trade of livestock and livestock products.* (Delft: IHE) pp 1-198

[14] Cheng G D 2004 *Virtual water: a new idea for the safety strategy of water resources. Society of ecological economics of China* (The 2004 academic annual conference of China eco economic society 2004)

# Monitoring lake level changes by altimetry in the arid region of Central Asia

**Y Zhao[1,2], J J Liao[1], G Z Shen[1] and X L Zhang[3]**

[1] Key Laboratory of Digital Earth Sciences, Institute of Remote Sensing and Digital Earth, Chinese Academy of Sciences, Beijing 100094, China;
[2] College of Resources and Environment, University of Chinese Academy of Sciences, Beijing 100049, China
[3] School of Earth Sciences and Resources, China University of Geosciences, Beijing 100083, China

Email: shengz@radi.ac.cn

**Abstract**. The study of lake level changes in arid region of Central Asia not only has important significance for the management and sustainable development of inland water resources, but also provides the basis for further study on the response of lakes to climate change and human activities. Therefore, in this paper, eleven typical lakes in Central Asia were observed. The lake edges were obtained through image interpretation using the quasi-synchronous MODIS image, and then water level information with long period (2002-2015) was acquired using ENVISAT/RA-2 and Cryosat-2 satellite borne radar altimeter data. The results show that these 11 lakes all have obvious seasonal changes of water level in a year with a high peak at different month. During 2002 – 2015, their water levels present decreased trend generally except Sarygamysh Lake, Alakol Lake and North Aral Sea. The alpine lakes are most stables, while open lakes' levels change the most violently and closed lakes change diversely among different lakes.

## 1. Introduction

Arid/semi-arid region of Central Asian with relatively dense distribution of lakes in the world is more than 88000 $km^2$ [1]. Lake is not only a sensitive indicator of climate and environment change, but also the indispensable foundation of human survival and social progress. Many studies have indicated that the inland lakes here have changed dramatically in recent decades, causing a series of ecological disasters [2, 3]. The water level is a significant indicator of lake change. Therefore observing lake level changes in a long time of the Central Asian is necessary.

Traditionally, the lake level monitoring is mainly through gauge data for a specific lake. However, these data can't be presented in the same reference system with inconsistent data accuracy. So, given large geographical span of Central Asia, it's difficult to use conventional means to obtain lake level changes of long time series throughout this region. Satellite altimetry makes lakes in remote area observable, which provides convenience for water level change study with large scale and long period. Satellite altimetry was used as an alternative tool to measure lake levels since the 1990s [4]. Different altimeters have been used along with its development, from Topex/Poseidon (T/P), ERS-1, ERS-2, ENVISAT to ICEsat, Jason-2 and Cryosat-2, with an accuracy ranging from centimeters to decimeters [5, 6, 7, 8, 9].

ENVISAT/RA-2 is currently the most widely used altimeter in monitoring lake level changes with higher time resolution, while Cryosat-2 makes the lake level observation using satellite data continuous. In this paper, the 11 typical inland lakes in Central Asia were studied, and water level information with long period (2002-2015) was acquired using ENVISAT/RA-2 and Cryosat-2 radar altimeter data. Then the characteristics of seasonal and inter-annual lake level variation were analyzed.

## 2. Study area and data

### *2.1. Study area*

As shown in Fig.1, the study area is the five countries in Central Asia located in the mid-latitudes of the north-western hemisphere (46°30′E – 87°30′E and 34°39′N – 56°16′N), including Kirgizia, Uzbekistan, Kazakhstan, Turkmenistan and Tajikistan. This region has a temperate continental climate with the average annual rainfall of 100 – 400 mm. Most areas are severe drought with little rain and large evaporation. This study selects 11 inland lakes as the research object.

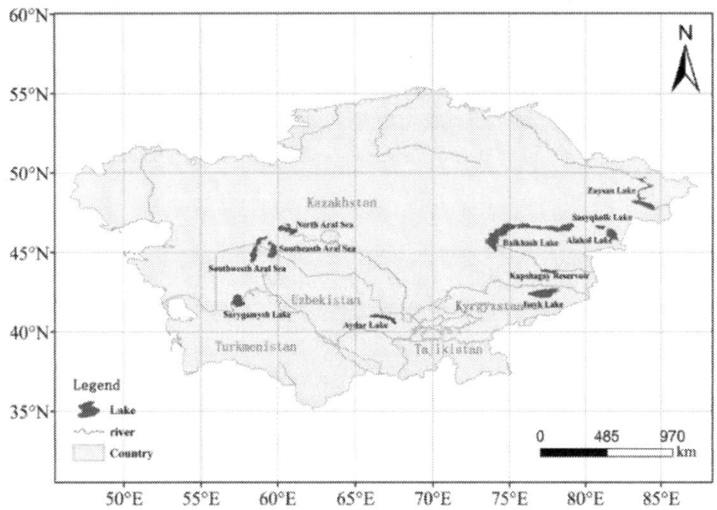


**Fig.1.** Map of study area and lakes distributions

### *2.2. Data*

In this paper, three datasets were used, including the image data of MODIS13Q1 from National Aeronautics and Space Administration (NASA), ENVISAT RA-2 and Cryosat-2 Geophysical Data Records (GDR) provided by the European Space Agency (ESA).

The ENVISAT RA-2 was launched in 2002 and stopped working in April 2012. It has been successfully used to monitor inland waters with accuracy about 0.3m [10, 11]. The RA-2 is a nadir-pointing instrument with a repeat cycle of 35 days. ENVISAT RA-2 GDRs provides four satellite-to-ground retracked ranges, including Ice-1, Ice-2, Ocean and Sea ice retracker. And Ice-1 was considered as the best data for lake level monitoring, which was used in this study [6, 12].

The Cryosat-2 satellite was launched in April 2010, carrying SIRAL altimeter, which is operating in three modes, including Low Resolution mode (LRM), SAR mode and SAR Interferometic (SARIn) mode. It has a 369-day repeat cycle with shifting sub-cycles of 30 days. The study area is mainly covered by LRM and SARIn observation mode. Furthermore, GDRs of LRM mode provides three different retrackers: UCL, refined CFI and refined OCOG, and the last one (OCOG) was used in this study.

## 3. Methods

To ensure that the data points are on the lake surface, boundary of each lake was firstly extracted by the method of density slice using NDVI (Normalized Difference Vegetation Index) image of MODIS13Q1 data, which were near the time altimeter visited.

Then, corresponding tracks of ENVISAT RA-2 passing each lake were selected. For example, there are 3 tracks (212, 425 and 969) overpassing Sarygamysh Lake before October 2010. Water Surface Height was calculated by subtracting the corrected range from the satellite altitude:

Water Surface Height = Altitude - Corrected Range

Where

Corrected Range = Range + Wet Troposphere Correction + Dry Troposphere Correction + Ionosphere Correction

The lake level was calculated above the geoid by the following formula:

Lake Level = Water Surface Height – Geoid - Solid Earth Tide Height - Pole Tide Height

Next, the data was further processed. **Step 1**, the points outside the boundary of the lake was deleted. **Step 2**, for all observed points of a lake, delete the obvious abnormal values using visual interpretation. **Step 3**, calculate the standard deviation ($\sigma$) and mean value (m) of those remaining footprints of a day, and remove those outliers which distance from m more than three times of $\sigma$ ($3\cdot\sigma$ criterion). Then calculate the mean value of a day as this day's water level. **Step 4**, delete the abnormal single day water level using $3\cdot\sigma$ criterion mentioned above. **Step 5**, get a relatively clean water level sequence using Gauss filter to remove noise for the whole day water level, with a filtering window of half a year [13]. **Step 6**: average the day water level of a month as the level of that month.

For Cryosat-2 data, the process is basically the same as ENVISAT RA-2 except two different points. First, the tracks passing a lake aren't fixed because of its shifting sub-cycles. Second, the Cryosat-2 GDRs provide geodetic height of points that have been corrected. So, lake level could be obtained by subtracting the Geoid from the given geodetic height.

Finally, combine the water level results derived from the two altimetry datasets to obtain water level time series of each lake from 2002 to 2015. Elevation system conversion is needn't in this study, because both the datasets use WGS84 & EGM96 reference system. To remove the remaining bias between them, we calculate the mean difference between them within the same time period. Then transform Cryosat-2 to the level of ENVISAT RA-2 by adding the difference.

## 4. Results

This study analyzed the characteristics of seasonal and inter-annual water level variations of 11 lakes in Central Asia. The results (Fig.2) show that they all have obvious seasonal changes of water level in a year with a high peak. Balkhash Lake has a relatively early water level high peak in about March, while Issyk Lake reaches the peak in September. Others have the maximum water level during May to July.

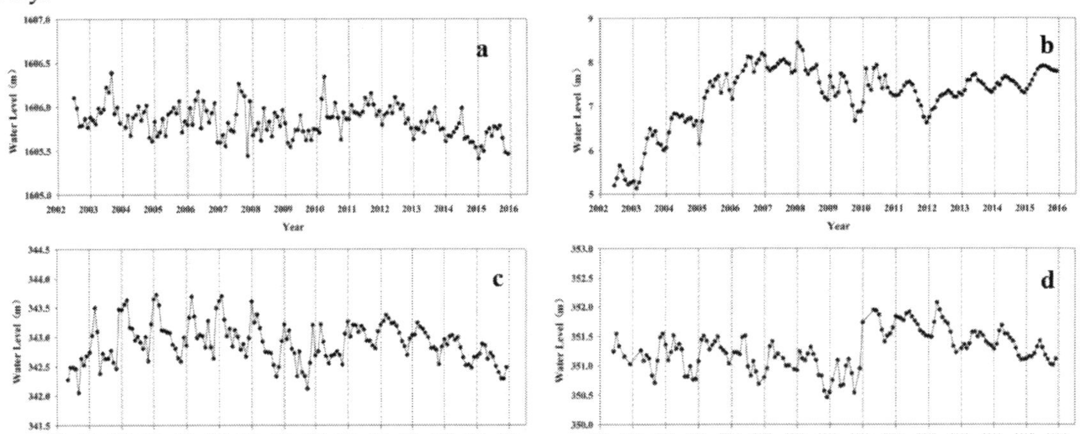

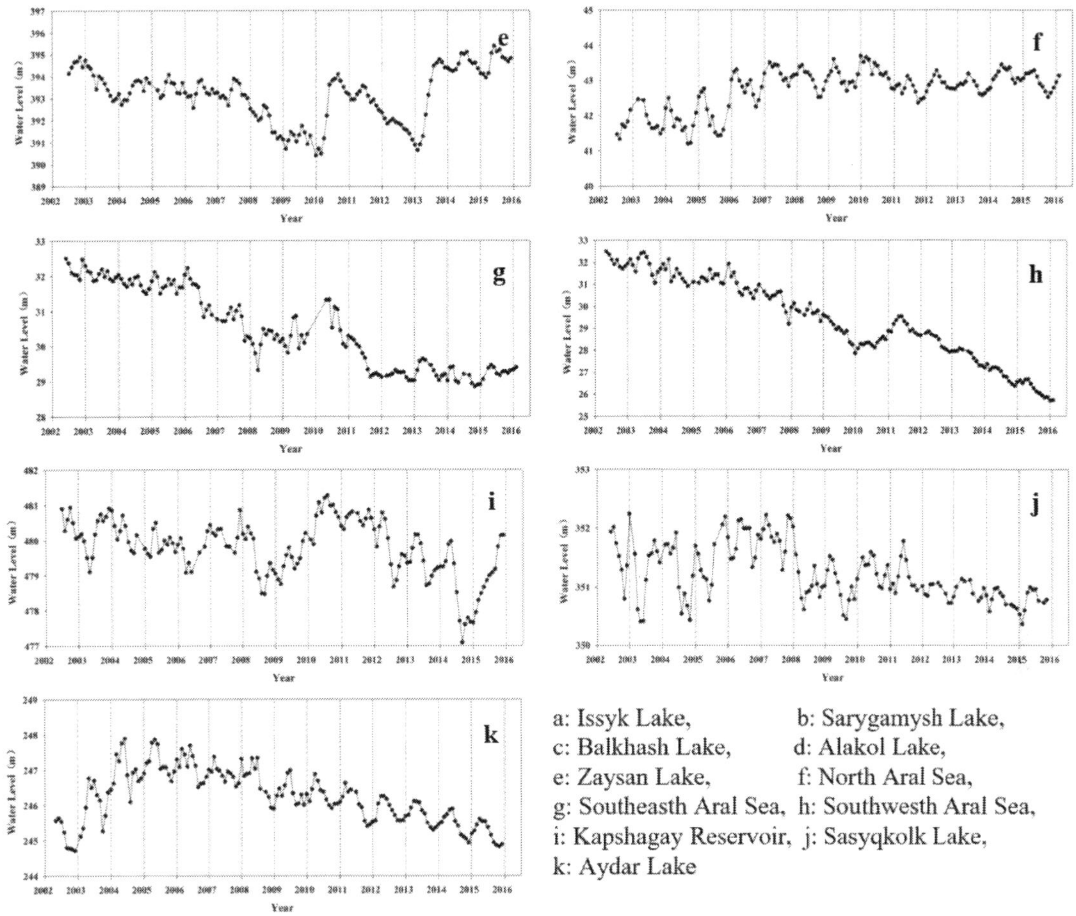


**Fig.2.** Water level time series of the 11 lakes from 2002 to 2015

We obtained the seasonal water level variations by calculating the mean difference between annual maximum and minimum value of water level (Fig.3). Zaysan Lake and Kapshagay Reservoir have the largest seasonal water level change, which is more than 1.2 meter, while Issyk Lake' is only about 0.5 meter.

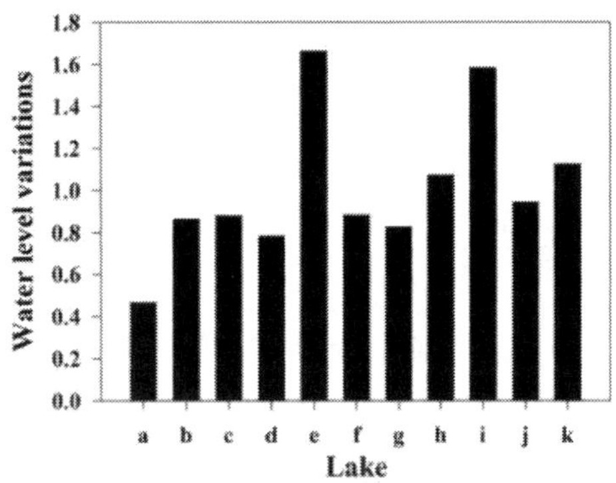


**Fig.3.** Seasonal water level variations (a-k the same as in Fig.2)

The mean temporal trends of lake level were calculated by a simple linear regression (Table.1). They show that Lake level generally decreases except three closed lakes among them: Sarygamysh Lake, Alakol Lake and North Aral Sea. Southeasth Aral Sea and Southwesth Aral Sea have the most significant continuing dropping change with high $R^2$ of 0.906 and 0.969 and the mean trend of -0.245 m/a and -0.450 m/a respectively.

**Table 1.** Statistics of lake level change information derived from ENVISAT RA-2 and Cryosat-2 Altimetry data

| Lake name | Longitude (°) | Latitude (°) | Start date (yyyymmdd) | End date (yyyymmdd) | Observation days | Trend (m/a) | $R^2$ | Types of lake |
|---|---|---|---|---|---|---|---|---|
| Issyk Lake | 77.3 | 42.4 | 20020714 | 20151227 | 687 | -0.014 | 0.267 | Alpine |
| Sarygamysh Lake | 57.4 | 41.9 | 20020616 | 20151208 | 315 | 0.111 | 0.389 | Closed |
| Balkhash Lake | 76.6 | 46.6 | 20020524 | 20151229 | 1825 | -0.003 | 0.004 | Closed |
| Alakol Lake | 81.7 | 46.1 | 20020612 | 20151208 | 258 | 0.025 | 0.149 | Closed |
| Zaysan Lake | 83.9 | 48.1 | 20020731 | 20151215 | 547 | -0.02 | 0.006 | Open |
| North Aral Sea | 60.6 | 46.5 | 20020727 | 20160221 | 367 | 0.091 | 0.538 | Closed |
| Southeasth Aral Sea | 59.7 | 45.1 | 20020626 | 20160225 | 401 | -0.245 | 0.906 | Closed |
| Southwesth Aral Sea | 58.5 | 45.2 | 20020525 | 20160227 | 362 | -0.45 | 0.969 | Closed |
| Kapshagay Reservoir | 77.5 | 43.8 | 20020711 | 20151218 | 325 | -0.087 | 0.333 | Open |
| Sasyqkolk Lake | 80.9 | 46.5 | 20020615 | 20151122 | 253 | -0.059 | 0.501 | Open |
| Aydar Lake | 66.8 | 40.9 | 20020526 | 20151222 | 558 | -0.074 | 0.195 | Open |

These 11 lakes are categorized into three types according to the type of basin and the source of the water supply: Alpine Lake, Closed Lake and Open Lake. As an alpine lake, Issyk Lake is the most stable one with the minimal seasonal and inter-annual water level changes. For the closed lakes, Balkhash Lake and Alakol Lake have a small change in general while they aren't as stable as alpine lakes. Alakol Lake's water level apparently increases in Month 2010. The water level of North Aral Sea and Sarygamysh Lake are rising, while Southeasth Aral Sea and Southwesth Aral Sea are dropping. North Aral Sea is tending towards stability after rising in 2006. Open lakes in arid region of Central Asia are decreasing generally. Their water levels changed most violently during 2002 – 2015, such as Kapshagay Reservoir, Zaysan Lake and Aydar Lake. Moreover, the water levels of open lakes in this area also have greater seasonal changes than others.

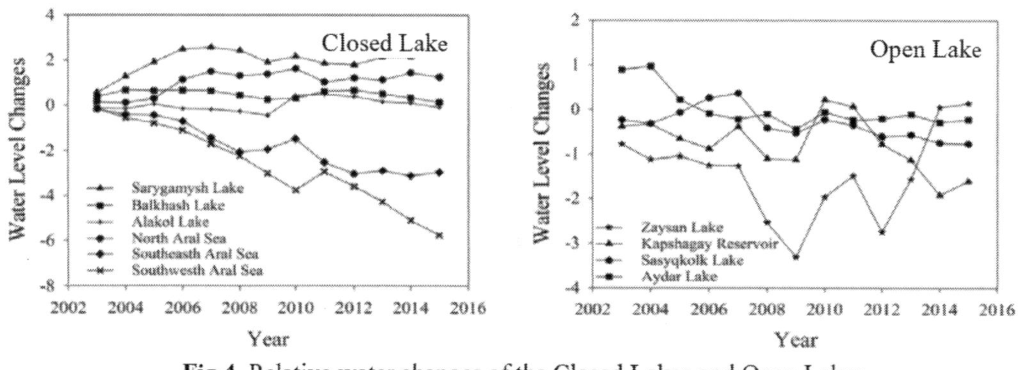


**Fig.4.** Relative water changes of the Closed Lakes and Open Lakes.

## 5. Conclusions

Monitoring water level using satellite altimeter, especially in a large scale and remote regions, is really meaningful. Water level time series of 11 main lakes in the arid region of Central Asia from 2002 – 2015 were derived from ENVISAT RA-2 and Cryosat-2 GDRs. The results show that:

(1) These 11 lakes all have obvious seasonal changes of water level in a year with a high peak at different month.

(2) These lakes' water levels show decreased trend generally except Sarygamysh Lake, Alakol Lake and North Aral Sea.

(3) The alpine lakes are most stables, while open lakes' levels change most violently and closed lakes change diversely among different lakes.

**Acknowledgments**
ENVISAT RA-2 and Cryosat-2 Altimetry data used in this study are provided by ESA. And MODIS image data is from NASA. This paper is supported by the National Key Research and Development Program of China (2016YFB0501501), the Product and Application of the Typical Scientific Factors of the "One-Three-Five" Project of Chinese Academy of Sciences (No. Y6SG0300CX), Natural Science Foundation of China (Grant No. 41401483), and Opening Fund of Key Laboratory of Poyang Lake Wetland and Watershed Research (Jiangxi Normal University), Ministry of Education (No. PK2015001). The author wishes to thank all above.

**References**

[1] Lehner B and Doll P 2004 Development and validation of a global database of lakes, reservoirs and wetlands *J. Hydrol.* **296** 1-22
[2] Wang X, Ding Y J, Liu S Y, Jiang L G, Wu K P, Jiang Z l and Guo W Q. 2003 Changes of glacial lakes and implications in Tian Shan, central Asia, based on remote sensing data from 1990 to 2010 *Environmental Reasearch Letters* **8** 575-91
[3] Bai J, Chen X, Li J, Yang L and Fang H. 2011 Changes in the area of inland lakes in arid regions of central Asia during the past 30 years *Environ. Monit. Assess* **178** 247-56
[4] Birkett C M. 1994 Radar altimetry: A new concept in monitoring lake level changes Trans., *Am. Geophys. Union* **75** 273-5
[5] Hwang C, Peng M F, Ning J, Luo J and Sui C H 2006 Lake level variations in China from TOPEX/Poseidon altimetry: data quality assessment and links to precipitation and ENSO *Geophys. J. Int.* **161** 1-11
[6] Frappart F, Calmant S, Cauhope M, Seyler F and Cazenave A 2006 Preliminary results of ENVISAT RA-2-derived water levels validation over the Amazon basin *Remote Sens. Environ.* **100** 252-264
[7] Fabrice P, Bala S, Pey R, Fabien D, Gopalakrishna V, Rahman A and Rossow W 2002 Ganga - Brahmaputra river discharge from Jason - 2 radar altimetry: An update to the long - term satellite - derived estimates of continental freshwater forcing flux into the Bay of Bengal *J. Geophys. Res.: Atmos.* **117** C11021
[8] Wang X W, Gong P, Zhao Y Y, Xu Y, Cheng X, Niu Z G, Luo Z C, Huang H B, Sun F D and Li X W 2013 Water-level changes in China's large lakes determined from ICESat/GLAS data *Remote Sens. Environ.* **132** 131-144
[9] Kleinherenbrink M, Lindenbergh R C and Ditmar P G 2015 Monitoring of lake level changes on the Tibetan Plateau and Tian Shan by retracking Cryosat SARIn waveforms *J. Hydrol.* **521** 119-131
[10] Silva J S D, Calmant S, Seyler F, Filho O C R, Cochonneau G and Mansur W J 2010 Water levels in the Amazon basin derived from the ERS 2 and ENVISAT radar altimetry missions *Remote Sens. Environ.* **114** 2160-81
[11] Jiang W P 2008 Water Level Variation of Qinghai Lake from Altimeteric Data *Geomatics & Information Sciences of Wuhan University* **33** 64-64 (In Chinese with English Abstract)
[12] Medina C E, Gomez-Enri J, Alonso J J and Villares P 2008 Water level fluctuations derived from ENVISAT Radar Altimeter (RA-2) and in-situ, measurements in a subtropical waterbody: Lake Izabal (Guatemala) *Remote Sens. Environ.* **112** 3604-17
[13] Gao Y G 2006 *Lake level variations from satellite altimetry* (Nanjing: Hohai University)

# Construction of a Distributed-network Digital Watershed Management System with B/S Techniques

**W C Zhang**[1], **Y M Liu and J Fang**[2]
[1]Key Laboratory of Digital Earth Science, Institute of Remote Sensing and Digital Earth (RADI), Chinese Academy of Sciences (CAS), Beijing 100094, P. R. China
[2] International Institute for Earth System Science (ESSI), Nanjing University, Nanjing 210093, P. R. China.

E-mail: zhangwc@radi.ac.cn

**Abstract.** Integrated watershed assessment tools for supporting land management and hydrologic research are becoming established tools in both basic and applied research. The core of these tools are mainly spatially distributed hydrologic models as they can provide a mechanism for investigating interactions among climate, topography, vegetation, and soil. However, the extensive data requirements and the difficult task of building input parameter files for driving these distributed models, have long been an obstacle to the timely and cost-effective use of such complex models by watershed managers and policy-makers. Recently, a web based geographic information system (GIS) tool to facilitate this process has been developed for a large watersheds of Jinghe and Weihe catchments located in the loess plateau of the Huanghe River basin in north-western China. A web-based GIS provides the framework within which spatially distributed data are collected and used to prepare model input files of these two watersheds and evaluate model results as well as to provide the various clients for watershed information inquiring, visualizing and assessment analysis. This Web-based Automated Geospatial Watershed Assessment GIS (WAGWA-GIS) tool uses widely available standardized spatial datasets that can be obtained via the internet oracle databank designed with association of Map Guide platform to develop input parameter files for online simulation at different spatial and temporal scales with Xing'anjiang and TOPMODEL that integrated with web-based digital watershed. WAGWA-GIS automates the process of transforming both digital data including remote sensing data, DEM, Land use/cover, soil digital maps and meteorological and hydrological station geo-location digital maps and text files containing meteorological and hydrological data obtained from stations of the watershed into hydrological models for online simulation and geo-spatial analysis and provides a visualization tool to help the user interpret results. The utility of WAGWA-GIS in jointing hydrologic and ecological investigations has been demonstrated on such diverse landscapes as Jinhe and Weihe watersheds, and will be extended to be utilized in the other watersheds in China step by step in coming years

## 1. Introduction

Watersheds are considered the most effective geographical unit for water management. Because of their topographical delineation, they include an (somewhat) independent drainage network of streams. However, watershed limits are frequently incongruent with administrative boundaries. Managing a transboundary watershed of this kind is often complication and requires what may be sparse resources of time, money and resources. The accurate depiction of earth surface processes and their responses to land cover, climate, or managerial change has been the goal of hydrologists for more than a century[1]. Spatially distributed models are by definition data-intensive, and if these models are to be applied on an operational basis, there is a critical need for automated PC-based procedures to store, access, and prepare data for modeling. GIS has been shown to be a valuable tool in providing watershed management solutions in the past. Geodatabases on soil structure, land use, vegetation and topography are paired with equations such as the Revised Universal Soil Loss Equation and models such as GIBSI (Gestion Intégrée des Bassins versants à l'aide d'un Système Informatisé) to provide decision makers with various management scenario outcomes.

This manuscript presents the Web-based Automated Geospatial Watershed Assessment GIS (WAGWAG) tool, a multipurpose hydrologic analysis system for use by watershed, natural resource, and land use managers and scientists in performing watershed- and basin-scale studies. It was developed under the following guidelines:

(1) Provide a simple, direct, and repeatable method for hydrologic model parameterization
(2) Use only basic, attainable GIS data
(3) Be compatible with other geospatial watershed-based environmental analysis software
(4) Be useful for scenario and alternative futures simulation work at multiple scales.

WAGWAG is an extension for the Environmental Systems Research Institute's ArcView versions 3.4, a widely used and relatively inexpensive PC-based GIS software package (trade names are mentioned solely for the purpose of providing specific information and do not imply recommendation or endorsement by the authors). The GIS framework is ideally suited for watershed-based analysis, which relies heavily on landscape information for both deriving model input and presenting model results from the other sources [2]. This facilitates comparative analyses of the results from multiple environmental assessments, thus making it particularly valuable for interdisciplinary studies, scenario development, and alternative futures simulation work. WAGWAG is distributed freely via the internet as a modular, open-source suite of programs (www.tucson.ars.ag.gov/agwa).

## 2. WebGIS general situation

The World Wide Web geographic information system is a new technology which expands and consummates geographic information system by using interactive network technology, and also it is a new technical method which unifies the geographic information system and the interactive net technology. People may conveniently gain on the Web browser for any kind of distributional geography space data and analysis from any WWW node through WebGIS. Its appearance enables GIS to move towards the society and realize GIS data-sharing, and it can serve for widespread social group. Because WebGIS is based on Internet and WWW environment and the distributed platform, it has several characteristics in the aspect of data management and the organization, such as supporting ultra-large data set, introducing large-scale database management system, picking-up the space data high quality, managing distributed data, uniting the data and system closely and so on[3] .

Presently, along with new technical and the hardware equipment renews, the application domain is widespread day by day. People's request for using information is also unceasing deepening and developmental. And these provide extremely broad prospects for development of application of WebGIS. For example, digital Earth involves main technologies such as computer, network correspondence, remote sensing, global positioning system, geographic information system as well as magnanimous data storage processing, image intelligence processing, database technology and so on, majority of these technologies may manifest in WebGIS. Similarly, the construction of digital watershed also utilizes these technologies, and WebGIS is the key to integrate these technologies.

## 3. Hydrology model for forecasting and dispatching management of watershed

Hydrology model for watershed is playing a vital role in researching hydrology rule and solving actual production problem [4]. About flood prevention and disaster decreased, the hydrology model is the core part of modern real-time flood forecasting and dispatching system. It is the key technology to increase the precision of forecast and the length of forecast period. About sustainable use of water resources, the hydrology model is rationale for appraisal, development, utilization and management of water resources. About water environment and ecosystem protection, the hydrology model is the main platform for constructing surface pollution model and ecology appraisal model. In addition, the hydrology model is an effective tool for analyzing and researching climatic change and the influence of human activity to the flood, the water resources, and the water environmental. At present, the distributed hydrology model which is tight bond based on physical foundation is the important research direction of hydrology simulation, and it already has obtained successful application in the water conservancy and correlation domain, just as the Xin'anjiang model.

The Xin'anjiang model is a conceptive basin rainfall runoff model. This model divides the entire basin into many unit basins, carries on runoff yield and concentration to each unit basin, and obtains the unit basin flow out rate. Then it takes flood routing of river course that exports- following and adds each unit basin flows together, at last obtains the basin total outflow. The core of the model is the stored-full runoff and the calculation module based on storage capacity curve of the basin. The model mainly adapts in moist and half moist area, it has high calculation accuracy and good application in hydrologic forecast work at home and abroad. The flow chart of Xin'anjiang (three water sources) model see Figure 1 [5]. In the chart, inputs are p as rains and EI as evaporating dish transpiration rate, outputs are Q as outlet section total flow and E as evapotranspiration. In the square is the state variables, outside the square is the model parameter.

The Weihe watershed has the great natural precipitation. This is dissimilar with other northwest local watershed. It is not belong to arid or semi-arid region, but is divided into half moist area. Therefore the Xin'anjiang model can carry on the hydrology simulation well on this watershed.

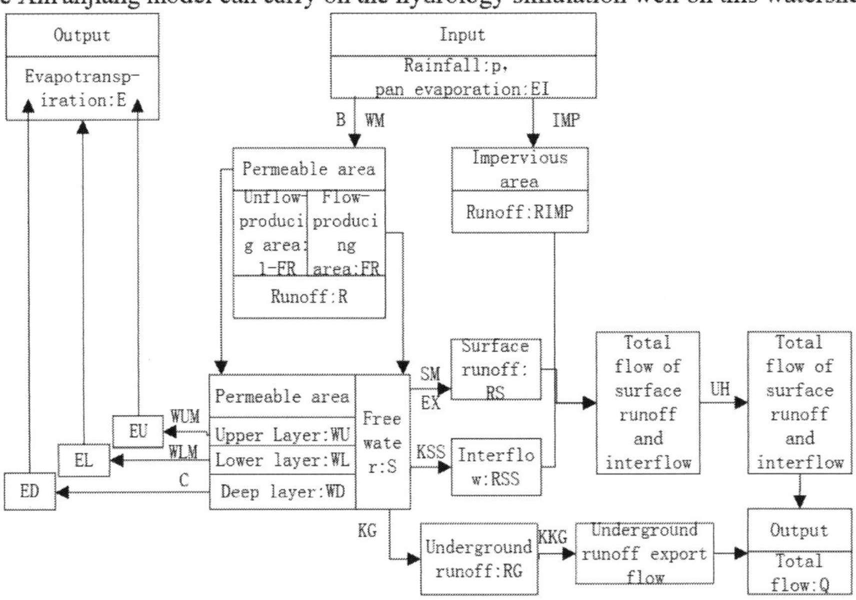


**Figure 1.** Flow chart of Xin'anjiang (three water sources) model

## 4. GIS data and distributed database

### *4.1. Characteristics of GIS data*

Sections should be numbered as follows: GIS has already stepped into the development stage that technical-driven changing to application-driven. The application of GIS in the project level and department level has obtained huge benefit, but when it penetrates into the process of commercialized and socialized, meets unprecedented difficult problem which is data-sharing. Data is the most expensive part in the GIS project. At present, we are facing the condition one is, there are massive GIS data resources along with the GIS widespread application; condition two is, GIS system was considered as the information isolated island, respective department always choices different GIS platform according to own special details and application demand, and this makes data-sharing difficult[6].

The distribution of GIS resources has the geography feature, special topic and object-oriented, therefore, the spatial database of GIS system should be had region, special and object-oriented laminations. The region lamination are the horizontal distribution of spatial data, this is decided by the regional characteristic of GIS resources. The special lamination is the vertical distribution of spatial database, this is decided by the vertical distribution of spatial database. The special lamination is usually advantageous discrimination for each different department's specialized data in identical physiographic region. The object-oriented lamination is regarded as classified the spatial data according to the attribute of itself, and it is usually divided into zero-dimensional, one-dimensional, two-dimensional and three-dimensional objects. The zero-dimensional object is point object, the one-dimensional object is line object, and the two-dimensional object is region object. For example, the departments' location is belong to point object; road and river is one-dimensional object; And the administrative division of township can be regarded as region object in a city geography system.

The digital Earth project and also its sub-project digital watershed are facing the question how to manage and use the magnanimous GIS data. Therefore, when constructing a WebGIS digital watershed system, the construction of database is the most important.

### *4.2. General situation of distributed database*

The distributed database system is a database that physical dispersion and logical collection. It uses the network to connect several logical units which are belonged to different geographical position but need some level degree control and management. They compose a unified database system together, and it is unification of distributed character and central tendency. The distributed database system has following characteristics [6]:

(1) Physical distribution: data stored in distributed database are not saves in one node, but in many nodes which connected by network;

(2) Logical integrity and visit transparency: data stored in distributed database system are physical dispersion in each node, but logically is actually in a whole, and can be visited transparently by any user;

(3) Node autonomy: data on each node managed by spatial database node itself, it has the autonomous handling ability, and completes application by itself;

(4) Certain data redundancy: The distributed spatial database system enhances usability and reliability of the system by certain redundancy mechanism, and it also can improve system performance.

The distributed database system has nimble architecture, and can adapt distributed management and control mechanism. The reliability of system is quite high, and the speed of response of partial application is quick. Also the expansibility is good; it may make nice data-sharing of different region and department. All these may decompose and disperse GIS data, and it can solve the problem of dispersed geographical location and massive quantity for GIS data, also it's easy to integrate a system that has efficient information-sharing mechanism.

## 5. The construction of digital watershed system

### *5.1. Data organization for digital watershed based on WebGIS*

The digital watershed involves many types of data, mainly includes map data, image data, measured data, text material data as well as multimedia data and so on, as shown in Figure 2.

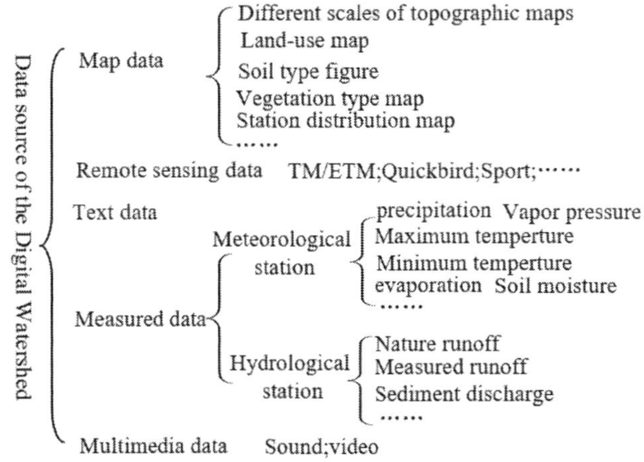


**Figure 2.** Data source of the digital watershed

The map data is a one of main data for digital watershed. It may make macroscopic and microscopic analysis in the application of digital watershed. The map used to establish the digital watershed is extremely widespread, and it besides each kind of national standards scale topographic map, the land-use chart, the soil type chart, the vegetation type chart, the weather station and hydrologic station distribution map and so on, all are the main map data. All these map data will be separately loaded to the digital watershed system as independent layers, and their attributes can be obtained through inquiry.

The image data is also an important data in digital watershed. It includes rich resources and environmental information, and it is one kind of large area, dynamic and near real-time data. It could be colored image synthesized by multi-wave band, also can be single wave band image that has been processed and has certain significance. But the remote sensing image is generally large, and it easy to lengthen reaction time in page's transmission process. Therefore, how to speed image data transmission in the network should be a question which is worth considering.

The data gained from detection, sampling and experiment have many characteristics such as accurately, real-time, dynamic, strong pertinence and so on. They are the main data sources which provide parameters for many kinds of models, and then the model can carry on hydrological simulation. For instance, there are many conventional meteorological data such as precipitation, maximum and minimum temperatures, evaporation, and also there are many conventional hydrological data such as runoff, sediment discharge and so on. The observation data from station involve extremely large data. For example, one watershed contains many sub-basins, each sub-basin contains many hydrological and meteorological stations, and each station contains many kinds of observation data each year and there are several or even several dozens years' data. So many data should be orderly organized with various attributes when stored into the digital watershed system. These observation data will follow the station map layers to load into the system, and the data can be extracted as records according to users' inquiry condition, and then it can generate report forms or graphs arbitrary. Moreover, during expanded application construction to the digital watershed, there are many relationships between these massive observation data as well as map data. So finding the relationship between data and data, data and layers will be helpful to construct digital watershed application-decision system.

Moreover, there are other kinds of data such as related text material, multimedia information and so on, and they are also component of the digital watershed. They include natural condition, social economy condition, correlation laws and regulations which are all related to the watershed. When establishing the digital watershed, it must make simple literal introduction as well as correlation explanation to this watershed, these may demonstrate on independent web pages. And it has the characteristics such as comprehensive, intuitionistic and so on.

This system database uses Oracle 9i.Oracle 9i is a large-scale distributed relational - object database which is based on high-level structured query language (SQL), namely fused object-oriented technical and Internet network application development technology in the relational database foundation. The large and middle scale management system, especially development of GIS and image information management system, requests large data processing with high speed. And Oracle database precisely has the characteristic. Its storage capacity is massive, and the data speed of data processing is quick. Meanwhile, Oracle is a cross-platform system, so it can provide many methods of exploitation that uses may choice the suitable one.

The spatial data in this system contain vector data and grid image data: Vector data are spatial data mostly being different ground objects or different types of thematic. Each kind of these space data store as a spatial table in database, and corresponding to the layers in Map Guide; it uses the ORDIMAGE object type which is in Oracle interMdia to save and manage the grid image data. Moreover, the observation data from weather station and hydrologic station will all be saved as general relational table.

### *5.2. The structure of digital watershed system*

This system takes Weihe watershed which located the intersection of Shanxi, Gansu and Ningxia as an example. It uses multi-layer distributed application model, and this may divide the system into data layer, logical layer and expression layer effectively. The biggest advantage of multi-layer structure designing is its expansive ability and load-balancing ability. Figure.3 is the system structure diagram.

This development language of network in this system is JSP (Java Server Page). It may produce dynamical and interactive Web server application procedure very conveniently, and could output colorful dynamic pages. Using Oracle for JDBC for visiting to Web page from database is very easy. It is extremely convenient for customers visiting database real-time, and so this realizes WebGIS visiting exterior data real-time. The technology of JSP usually unifies with Java Servlet, and it may insert Java code section and transfer exterior Java module to mark languages such as HTML and XML. Also it may use JavaBeans and EJBs Enterprise as front processing tools. JavaBeans realizes complex commercial logic and dynamic function perfectly, and the hydrology model inserted in this system just utilizes JavaBeans.

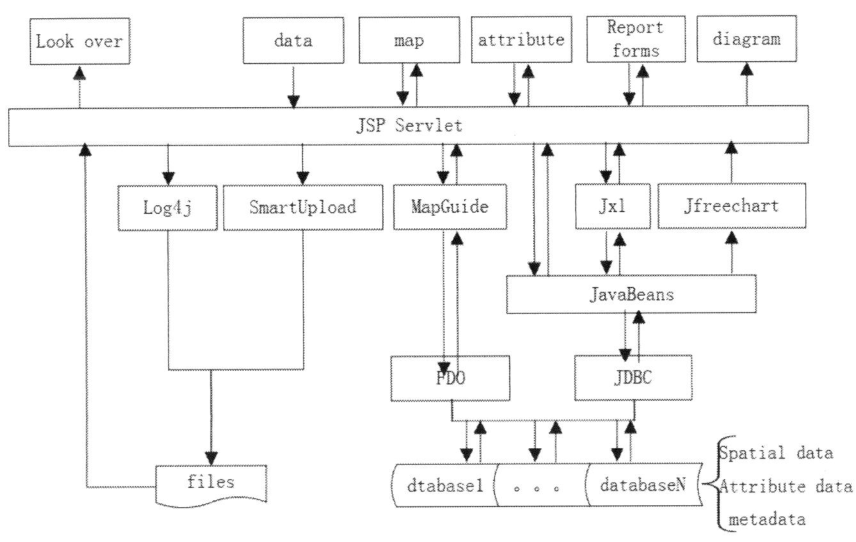


**Figure 3.** System structure diagram

### *5.3. The function structure of digital watershed*

This system realizes functions through concrete function module, and different module corresponding to different user class. There are two kinds of user in this system, one is common clients and the other is administrators of the watershed. The achievement demonstration part in the common client module is the watershed information issue system, it has independent page system. Users should register and be judged the jurisdiction rank in the website, and then it will demonstrate partial or complete map contents to the user. There are map-operation and the map-inquire for this sub-module's function design. Map-operation function contains map cruise, zoom-in and zoom-out, point selection, measure, buffer analysis and so on; Map inquiry function, not only realize space and attribute bidirectional inquiry, but also increase dynamic link, and it may click to look over detailed situation of each meteorological or hydrologic station, including basic information, picture introduction, station material inquiry, report and graph production and so on (home page of the website see Figure 4). The module designed for common clients see Figure 5 to show.

**Figure 4.** Home page of the website

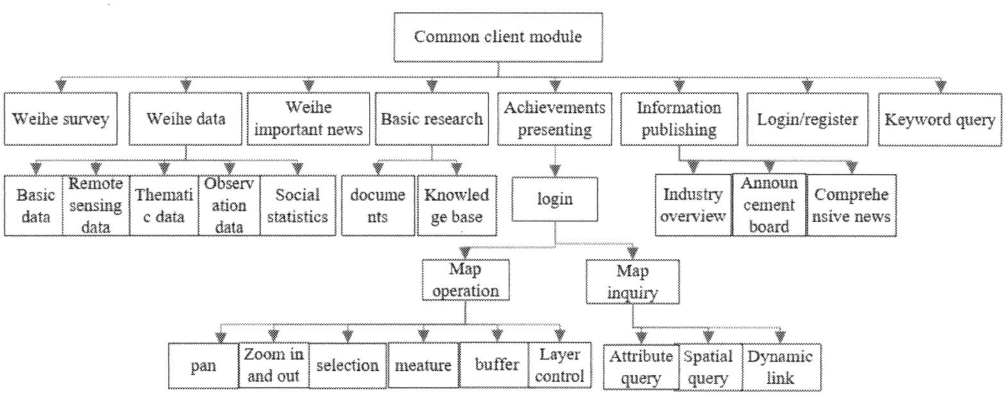


**Figure 5.** Module designed for common clients

The superintendent module besides has all functions of common clients module, but also increases database operation to manage users, homepage information as well as station material in the watershed system. Moreover, it also has all function of the achievement demonstrate part, and besides map operation and map inquiry. But also increase hydrology simulation function sub-module foundation for the superintendent module (take Xin'anjiang model as example). It codes the hydrology model and loads services to the web. Then hydrology simulation work will be carried on some sub-basin by using network database or users' local data. The module designed for Admin of the webpage is seen Figure 6, and the effects of digital Weihe website while WebGIS operating see Figure 7 and Figure 8.

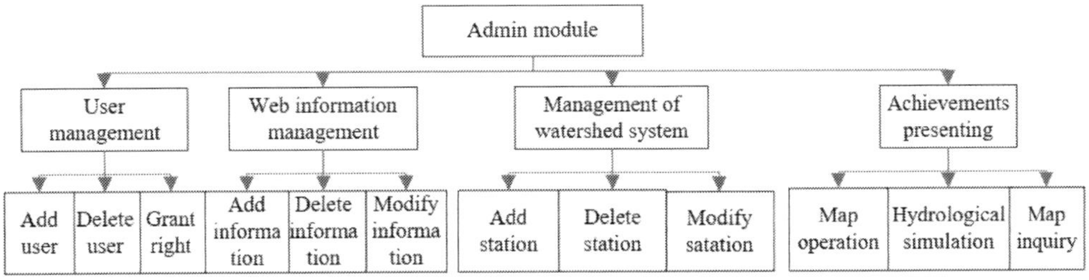


**Figure 6.** Module designed for admin of the webpage

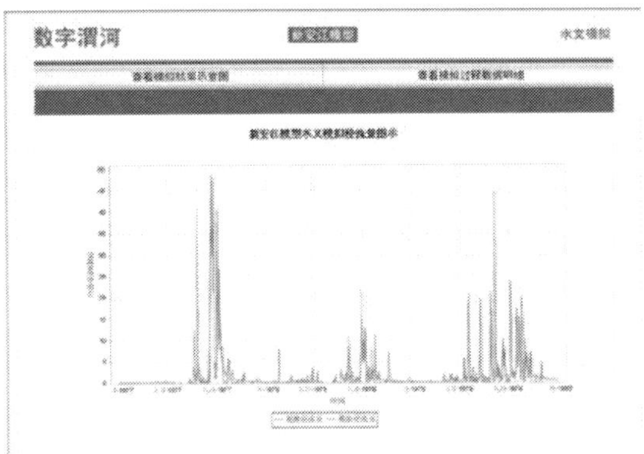


**Figure 7.** Distance measure

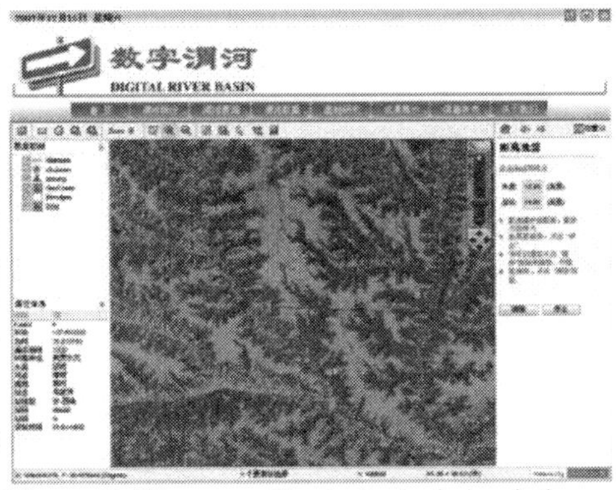


**Figure 8.** Diagram of hydrological simulating

## 6. Analysis of system performance optimization

Along with the development of WebGIS, users' requirements for data browsing, inquiry and analysis are higher and higher. Therefore, enhancing service performance and satisfying users' demand becomes the urgent matter in development of WebGIS. This part ponders and analyzes how to enhance the system performance effectively from two aspects which are distributed database construction of data and inquiry optimization.

The performance of WebGIS for digital watershed system is limited while transmitting mass map data. From the database design angle, besides the technology which has used, also can optimize system by changing database construction of data to enhance system performance. In this system, spatial data is divided into certain special layers according to the characteristics of spatial objects, each layer saves as a data sheet in database. Thus, a traditional map decomposes several special tables superimposition. And it can reduce the map exaggeration time, moreover because the map contains sole entity element, each table occupies less space. On the other hand, we usually don't use all the layers one time, only transfer some or several layers that actual need, and thus system resources occupied and consumption time when transmitting will have big improvement. Moreover, in the grid image data memory aspect, block storage, highly effective compression and pyramid structure technology, enables mass and high clear images have efficient demonstration and high transmission like general images.

If inquiry for not only one data server in the distributed database system (called this inquiry for overall situation inquiry), it must carry on optimize and decompose the overall situation inquiry. The inquiry optimization of distributed database system has two kinds of different goals: One kind of goal is minimum total cost takes as standard; the other kind of goal is short inquiry response time take for each as standard. Of specifically speaking, generally in the long-distance correspondence network, the speed of data transmission among stations is slower many times to transmission between memory and floppy disk in the single plane situation. Therefore, in this kind of distributed inquiry environment, the inquiry response time compare to the time communication needs can be ignored, and reducing the correspondence expense becomes the essential target. In the high speed local area network, the transmission time must be much shorter than the response time. In this case, the response time takes optimized goal [8].In this system, because of the big jurisdiction difference among client side users, and the jurisdiction indicates to a certain extent how many data may gain. In control department, it takes data bank administration and data maintenance, therefore first should consider the question of response time; while major part of common client are long-distance correspondence network client side, its jurisdiction has graduation limit, therefore it should take minimum total cost, namely reducing the correspondence expense is the just standard.

## 7. Conclusion

The digital watershed is one kind of multi-level structure, and it has widespread application prospect. Its integrity and systematic total view will bring brand-new situation of watershed management like flood-control to reduce disaster and water resources regulation. Because of many superiorities of WebGIS, the watershed GIS developing mainstream will certainly to be watershed WebGIS. Open style solution, lower and lower construction cost, more and more intellectualized construction method, WebGIS will realize processing of spatial data and data sharing maximum convenient and effective. Watershed WebGIS will be certainly to impel environmental protection, resources development, traveling development and electricity energy development of the watershed enormously.

**Acknowledgements**

This study was financially supported by the National Key Research and Development Program of China (Grant No.2016YFA0602302 and No.2016YFB0502502).

## References

[1] Xu G H 1999 “The Digital Earth” Construction, Promotes Sustainable Development of Chinese and Global *Remote Sensing Information* **4** 10-12 (In Chinese with English Abstract)

[2] Zhang Y C and Wang C 2001 Digital Watershed - An Important Region Level of Digital Earth *Water and Electricity Energy Science* **9** 1-3 (In Chinese with English Abstract)

[3] Li B X 2008 Research and Application of Web Geographic Information System (WebGIS) *Mapping and Spatial Geography Information* **4** 113-18 (In Chinese with English Abstract)

[4] Dong F F 2007 *Comparison of Xin'anjiang Model and TOPMODEL and Basic Frame of Flood Forecasting System* (Nanjing: Hohai University) (In Chinese with English Abstract)

[5] Zhai J R 1995 *Commonly Used Hydrologic Forecast Algorithm and Computational Procedure* (Zhengzhou: Zhengzhou Yellow River Water Publishing house) p 23 (In Chinese with English Abstract)

[6] Yi X F 2005 *Research and Application of Distributed Spatial Database Integration Engineering Which is based on WebGIS System* (Nanjing: Wuhan University of Technology) (In Chinese with English Abstract)

# Wide area lithologic mapping with ASTER thermal infrared data: Case studies for the regions in/around the Pamir Mountains and the Tarim basin

**Yoshiki Ninomiya[1] and Bihong Fu[2]**
[1] Geological Survey of Japan, AIST, Tsukuba, Japan
[2] Institute of Remote Sensing and Digital Earth, CAS, Beijing, China

E-mail: yoshiki.ninomiya@aist.go.jp

**Abstract.** After the authors have proposed the mineralogical indices, e.g., Quartz Index (QI), Carbonate Index (CI), Mafic Index (MI) for ASTER thermal infrared (TIR) data, many articles have been applied the indices for the geological case studies and proved to be robust in extracting geological information at the local scale. The authors also have developed a system for producing the regional map with the indices, which needs mosaicking of many scenes considering the relatively narrow spatial coverage of each ASTER scene. The system executes the procedures very efficiently to find ASTER data covering a wide target area in the vast and expanding ASTER data archive. Then the searched ASTER data are conditioned, prioritized, and the indices are calculated before finally mosaicking the imagery. Here in this paper, we will present two case studies of the regional lithologic and mineralogic mapping of the indices covering very wide regions in and around the Pamir Mountains and the Tarim basin. The characteristic features of the indices related to geology are analysed, interpreted and discussed.

## 1. Study area

Two study areas in the central Asia are examined: 1) The Pamir Mountains in Tajikistan and the surrounding area, bounded by latitude 37° to 40° North and longitude 71° to 74° East; 2) The Tarim basin and the surrounding area, bounded by latitude 36° to 42° North and longitude 78° to 90° East. The latter is quite vast, therefore, the lithologic mapping just for the north-eastern marginal area of the Tarim basin, bounded by latitude 39° to 42° North and longitude 87° to 90° East is shown here.

## 2. ASTER data, mineralogical indices and mapping procedures

ASTER was launched onboard Terra, the first of NASA's EOS series of satellites, in December 1999. It obtains multispectral image of the Earth not only in the VNIR and SWIR, but also in the TIR (five bands between 8 and 12-μm, 90-m resolution, NEΔT<0.3 K at 300 K). After the launch of ASTER in 1999, the ASTER VNIR and TIR subsystems keep observing the earth for more than 16 years, although the SWIR subsystem had stopped operation by trouble in 2008. ASTER data are available at no cost for everyone from the data distributor, the National Institute of Advanced Industrial Science and Technology (AIST) and USGS as a result of policy change announced in April 2016,

After the confirmation of analogies among the thermal infrared spectral properties measured in laboratory, in situ and remotely sensed ones [1], mineralogical indices, Quartz index (QI), carbonate

[1] correspondence author

index (CI) and mafic index (MI) are proposed [2] with the calculation among the ASTER-TIR bands:

$$QI = band11 \times band11 / band10 / band12, \quad (1)$$
$$CI = band13 / band14, \quad (2)$$
$$MI = band12 / band13 / CI^3 = band12 \times band14^3 / band13^4. \quad (3)$$

QI is positively sensitive to quartz, especially to the sedimentary rocks rich in quartz. Also, QI indicates quite low value for the sulphate minerals typically gypsum. Weaker than gypsum but feldspar minerals and granitic rocks rich in the feldspar minerals also indicates low value in QI. CI indicates high value for the typical carbonate rocks, i.e., limestone and dolomite. MI is inversely correlates to the bulk $SiO_2$ content of the target silicate or igneous rocks. MI indicates quite high value for the ultramafic rocks, the main composition of ophiolite. MI indicates quite low value for high silica rocks typically sedimentary rocks rich in quartz where QI indicates quite high. Many studies applying the indices have clarified the characteristics of the indices [3] as summarized partly in Table 1.

**Table 1.** Characteristic features of the indices for geological materials

| Expected material | Features in indices | Colour in index image |
|---|---|---|
| Quartz rich / Silicates poor with some carbonates | QI>1.05, MI<0.8 and CI>1.02 | Yellowish red |
| Quartz rich / Silicates poor with minor carbonates | QI>1.05, M<0.8 and CI<1.02 | Pure red |
| Quartz rich / Feldspars poor with some mafic minerals | QI > 1.05 and MI > 0.82 | Red to purplish red |
| Carbonates | CI > 1.05 | Light green |
| Sulphates | QI < 0.98 | Deep green |
| Ultramafic | MI > 0.92 | Light purple |
| Feldspars rich granitic rocks | Relatively low QI / Image pattern | Dark bluish green |
| Feldspars poor granitic rocks | Image pattern | Dark pinkish purple |
| Mafic rocks | Relatively high MI (≈0.90) | Deep blue |

Lithologic mapping are accomplished according to the characteristics of the indices summarized in Table 1. The ASTER radiance at the sensor data (Level-1B data) are used for the mapping.

The procedures for producing the mosaic lithologic maps covering wide area [3] are described as follows: Firstly, the browse images of the ASTER data covering the required areas are checked before downloading. The next step is to calculate the indices (i.e., QI, CI and MI) for the respective data, and carefully check the images of the VNIR data and the indices manually. The CI index images should be checked carefully as they can be heavily affected by the atmospheric water content and as a result should be omitted as useless data. During the course of the checking, the priority ranking of each ASTER data is determined, and the mosaic images of the indices covering the area of interest are produced as the regional maps. In the case of this study, we produce the mosaic images mapped for every 1°×1° region of the Earth surface, and then joined them together to produce the regional maps covering the entire study area.

## 3. Result

Figure 1 shows one of the result regional maps of the indices, the false colour composite image of the indices assigning QI for red, CI for green and MI for blue, for the Tajikistan Pamir Mountains study area. Figure 2 shows the same kind of the result for the north-eastern marginal area of the Tarim basin study area. For both of the cases, the fixed data range of each index, 0.97 to 1.055 for QI, 1.005 to 1.055 for CI and 0.79 to 0.95 for MI, is linearly converted to the intensity of each colour, so that the meaning on the colours or the intensities in the map images is kept through the cases. The boundary between the ASTER scenes is not so significant in the maps even though the fixed data range is applied, which suggests the stability of the indices even though the radiance at the sensor data without atmospheric corrections are adopted for calculating indices with the equations (1) to (3).

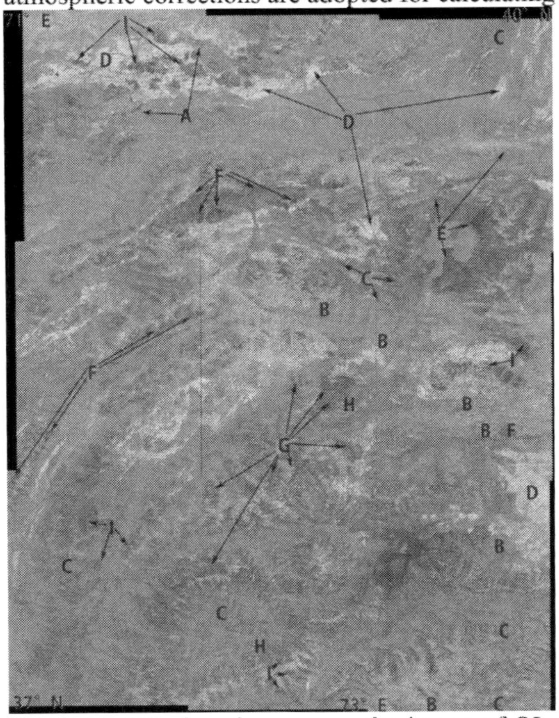

**Figure 1.** Mosaic colour composite image of QI, CI and MI with a linear contrast stretch of 0.97-1.055 (QI; Red), 1.005-1.055 (CI; Green) and 0.79-0.95 (MI; Blue) using ASTER Level-1B products, covering the Tajikistan Pamir Mountains study area bounded by latitude 37° to 40° North and longitude 71° to 74° East, in a Universal Transverse Mercator (UTM) map projection. The annotations indicate the locations discussed in the text.

**Figure 2.** Mosaic colour composite image of QI, CI and MI with a linear contrast stretch of 0.97-1.055 (QI; Red), 1.005-1.055 (CI; Green) and 0.79-0.95 (MI; Blue) using ASTER Level-1B products, covering the north-eastern marginal part of the Tarim basin study area bounded by latitude 39° to 42° North and longitude 87° to 90° East, in a Universal Transverse Mercator (UTM) map projection. The annotations indicate the locations discussed in the text.

## 4. Discussion

The lithologic features characterized with the indices are analysed based on Table 1. The prominent features are annotated with the alphabetic characters (a part of them are accompanied with the arrowed lines) in Figures 1 and 2, which indicate the geological bodies discussed below. The geological interpretations made here are just only with the features appeared on the indices without any presupposed geological information. In the sense, the discussions made here are preliminary ones, which need to be discussed further with the field geological data.

*4.1. The Tajikistan Pamir Mountains study area*

Ninomiya and Fu [4] made a lithologic mapping with the indices for the Chinese region of the Pamir Mountains. This study makes it for the Tajikistan Pamir Mountains area. Various kinds of lithologic units exhibit particular features in the colour index image (Figure 1). The bodies “A” denote yellowish red, expected as “Quartz rich / Silicates poor with some carbonates”. This type is predicted as non-metamorphosed quartzose sedimentary rocks. Quartzose rocks with reddish colour (QI > 1.05) are widely distributed; however, most of them in the Pamirs are also relatively high in MI (> 0.82). The bodies “B” with pure reddish colour are expected as “Quartz rich / Feldspars poor with some mafic minerals”, which are predicted as low-grade metamorphosed quartzose sedimentary rocks typically pelitic schist. The bodies “C” with reddish pink colour are with higher MI (> 0.85), which are predicted as high-grade metamorphic quartzose rocks namely gneissose sedimentary rocks.

The bodies “D” with light greenish colour are carbonate rocks. Also, the bodies “E” with deep greenish colour are gypsum deposits and layers. The bodies “F” of light bluish purple with high MI (> 0.92) are ultramafic rocks. The bluish bodies around the ultramafic rocks and the ones suggested as “I” are expected to be mafic rocks (e.g., gabbro and basalt). The dark greenish bodies “G” with characteristic texture including the ones around the snowy mountains seem “Feldspars rich granitic rocks”, and the ones of dark pinkish purple indicated as “H” seem “Feldspars poor granitic rocks”.

*4.2. The Tarim basin study area*

Various kinds of geological units exhibit particular features in the colour index image covering the Tarim basin study area (Figure 2). Outcrops are developed well in the northern part and the southernmost-eastern part. The other parts are covered with the deposits in the north-eastern part of the Tarim basin. The annotated capital characters indicate rocks, and small characters indicate deposits.

Many quartzose rocks with reddish colour are distributed for example the bodies “A”. In this area, most of the quartzose rocks are with relatively low MI (< 0.8). This suggests that they are non-metamorphosed sedimentary rocks. Also, many marine sedimentary layers of carbonates with light greenish colour are widely distributed for example the bodies indicated as “D”. The deep or dark greenish bodies “E?G?” are also distributed widely in the northern mountainous area. They are distributed with the relation to the bluish silicate rock bodies (typically the bodies “I”), so that perhaps they are the “Feldspar rich granitic rocks”. However, it needs to check in the field. The bodies “I” seem to be mafic rocks, and the ones “F” seem to be mafic to ultramafic rocks.

The deposits also denote the characteristic colours in the composite index image (Figure 2) here at the eastern marginal region of the Tarim Basin. The reddish colour of the bodies “a” suggests the high quartz content. The light greenish colour of the bodies “d” suggests the high calcite content. The deep greenish colour of the bodies “e” is reflecting the high gypsum content. The source of the reddish purple colour at the bodies “?” is not determined yet. The whitish yellow or whitish blue colour at the bodies “s” , which are distributed at both the deposit and mountainous regions, is reflecting the features of high for all the three indices (i.e., QI, CI and MI). Such features are expected for salt deposits. Comparing the multi-temporal images, most of these whitish yellow or blue bodies are not stable, that is, they are appearing or disappearing along with the condition of climate.

## References

[1] Ninomiya Y et al. 1997 A comparison of thermal infrared emissivity spectra measured in situ, in the laboratory, and derived from thermal infrared multispectral scanner (TIMS) data in Cuprite, Nevada, USA. *Int. J. Remote Sens.* **18** 1571-1581

[2] Ninomiya Y and Fu B 2002 Quartz index, carbonate index and SiO2 content index defined for ASTER TIR data. *Japan. J. Remote Sens. Soc.* **22** 50-61

[3] Ninomiya Y and Fu B 2016 Regional Lithological Mapping Using ASTER-TIR data: Case Study for the Tibetan Plateau and the Surrounding Area. *Geosciences* **6** 39

[4] Ninomiya Y and Fu B 2003 Extracting lithologic information from ASTER multispectral thermal infrared data in the northeastern Pamirs. *Xinjiang Geology* **21** 22-30

# Assessing soil erosion using USLE model and MODIS data in the Guangdong, China

**Feng Gao**[1, 2], **Yunpeng Wang**[1*], **Jingxue Yang**[3]
1 Guangzhou Institute of Geochemistry, Chinese Academy of Sciences, Guangzhou 510640, China
2 University of Chinese Academy of Sciences, Beijing 100049, China
3 Guangdong Research Institute of Water Resources and Hydropower, 510610, China

E-mail: wangyp@gig.ac.cn

**Abstract.** In this study, soil erosion in the Guangdong, China during 2012 was quantitatively assessed using Universal Soil Loss Equation (USLE). The parameters of the model were calculated using GIS and MODIS data. The spatial distribution of the average annual soil loss on grid basis was mapped. The estimated average annual soil erosion in Guangdong in 2012 is about 2294.47t/ ($km^2$.a). Four high sensitive area of soil erosion in Guangdong in 2012 was found. The key factors of these four high sensitive areas of soil erosion were significantly contributed to the land cover types, rainfall and Economic development and human activities.

## 1. Introduction

Soil erosion, as a global environmental crisis in the world, has attracted more attention in the developing countries. The Universal Soil Loss Equation (USLE), because it has a simple form, parameters can be easily obtained and also suitable for material shortage areas, has been widely applied to assess soil erosion in many nations since 1960s. As an empirical model, USLE serves to estimate annual soil loss. Many studies have demonstrated that USLE model can be used to predict effectively soil loss in different scales around the world [1-2]. Therefore, we select USLE model to estimate the annual rate of soil erosion in Guangdong Province, China.

USLE model needs information related to climate, soil type, topography, land use and landform to predict soil loss. In order to obtain these environmental variables effectively, integration of USLE, RS and GIS are often used [2]. Remote sensing data and GIS could provide accurate, timely, and real-time information on various aspects of the study area such as land use, soil distribution, vegetation coverage, etc. In this study, we used MODIS land use/cover product and MODIS NDVI product to evaluate the parameters (C and P factor) of the USLE model. The aim of this study was to determine the soil erosion variables associated with USLE model and spatial distribution of soil loss and evaluate the application of GIS, RS and USLE to determine average annual soil loss.

## 2. Materials and method

### *2.1 Study area and data source*

The study area, Guangdong Province, is located in the southern China. It lies between E 109° 45′ and E117° 20′ and N20° 09′ and N25° 31′ with an elevation ranges from 0 to 1876 m above

* Corresponding author.

mean sea level. It covers an area of approximately 179,800 km$^2$ and consists of highlands in eastern, western and northern borders hills and terraces in the middle and plains in the south. Guangdong enjoys a humid subtropical climate with abundant rainfall. The annual average temperature is between 18 ℃ and 24℃,and the annual mean precipitation reaches 1,700 mm. Winters are short, mild and relatively dry, while summers are long, hot and very wet. In addition, the vegetation coverage in this area is relatively higher than that in the northern China.

The data used in this study mainly include MODIS 500 m land use/cover product and MODIS vegetation indices acquired in 2012. The pedological map of Guangdong with a scale of 1:1,000,000 is a government issued atlas; digital elevation model (DEM) data are from the 90-m Shuttle Rader Topography Mission data of NASA; Daily rainfall data from 1961 to 2012 of the study area are downloaded from the China Meteorological Data Sharing Service System (http://cdc.cma.gov.cn/).

*2.2 Method*

In this study, the average annual soil loss was calculated using the USLE model, which is defined as:

$$A = R \cdot K \cdot L \cdot S \cdot C \cdot P$$

Where $A$ is the average annual soil loss in t/(hm$^2$*a); $R$ is rainfall erosivity factor (MJ mm hm$^{-2}$ h$^{-1}$ a$^{-1}$); $K$ is soil erodibility factor (t h MJ$^{-1}$ mm$^{-1}$); $L$ is slope length factor; $S$ is slope steepness factor; $C$ is cover and management factor, and $P$ is the support practice factor. The definition and method of calculation of each factor are laid out in the following sections.

*2.2.1 Rainfall erosivity factor (R)*

Rainfall erosivity is the impact factor of quantitative assessment of soil erosion. It reflects the ability of soil erosion caused by rainfall, which is closely related to rainfall, rainfall energy and rainfall duration. Wischmeier (1965) defined R as the product of rainfall energy and rainfall intensity of maximum 30 min. However, there is no detailed record of rainfall data in weather station in China. Chinese scholars proposed some simple approach to calculate rainfall erosivity. In this study, we used a simple calculation method of R value based on daily rainfall data proposed by Zhang (2002) [3]. Daily precipitation data from 1961 to 2012 collected from over 50 weather stations in Guangdong were used to calculate the mean R value of each station. The average regional R value was then calculated using kriging interpolation in ArcGIS 10.1 to acquire the spatial distribution of R factor.

*2.2.2 Soil erodability factor (K)*

Soil erodability is an important indicator that is used to evaluate the sensitivity of soil from erosion. It reflects the ability of soil detachment and transport by raindrop impact and surface flow. In this study, we use the calculation method of K value presented in EPIC [4] using the soil mechanical composition data derived from the second national soil survey in China.

*2.2.3 Topographic factor (LS)*

Topographic factor (LS) is the product of slope length factor (L) and slope steepness/slope gradient factor (S). Slope length is defined as the distance from the point of origin of overland flow to the point where either the slope gradient decreases enough that deposition begins or the runoff water enters a well-defined channel. Slope length factor (L) is the ratio of soil loss from the given length of slope to that from land having 22.13 m length of slope. The factor (L) from McCool [5] described as follows:

$$L = (\lambda / 22.13)^m \quad (1)$$

Where $\lambda$ is slope length (in meter), $m$ is an exponent with a value between 0.2 and 0. 5.

The application of $S$ is limited to slopes below 18% because it is the upper bound in USLE. We used USLE for slopes below 18% and the equation proposed by Liu (1994) for steeper ones [6].

*2.2.4 Cover management factor (C)*

The C factor measures the combined effect of all interrelated cover and management variables that can be easily changed to reduce erosion. It was proven that the C factor is closely related to vegetation cover. Thus, we used MODIS NDVI product to produce vegetation cover map for 2012 with 500 m resolution. The relationship between C factor and vegetation cover can be found in Ma (2002) [7].

*2.2.5 Support practice factor (P)*

Support practice factor is basically the ratio of soil loss from a land having specified conservation practices to that from a land ploughed in a direction parallel to the slope if all other conditions remain unchanged. These practices affect erosion by modifying the flow pattern, grade or direction of the surface runoff and by reducing the amount and rate of runoff. Generally, support practice factor is determined with help of land use land cover map. In this study, we used MODIS land use land cover product and related literature [7-8] to produce the spatial distribution of P value in the study area.

## 3. Results and discussion

The spatial distribution of average annual soil loss in the study area was calculated by multiplying cell-based USLE parameters in the specified 500 m cells for the year 2012(Figure 1). The average annual soil losses were divided into 6 classes: low erosion, mild erosion, moderate erosion, serious erosion, very serious erosion, and severe erosion. Various statistic indictors of different soil erosion intensity in 2012 were presented in Table 1. From Table 1, we can see that the average annual total erosion is $4.04\times10^8$t and the average annual erosion is 2294.47t/ ($km^2$.a) in Guangdong in 2012, which demonstrates that the whole performance of the intensity of soil erosion in Guangdong is low erosion. The soil erosion intensity in the study area is mainly focus on low erosion and mild erosion, with an area of 57.62% and 22.11%, respectively.

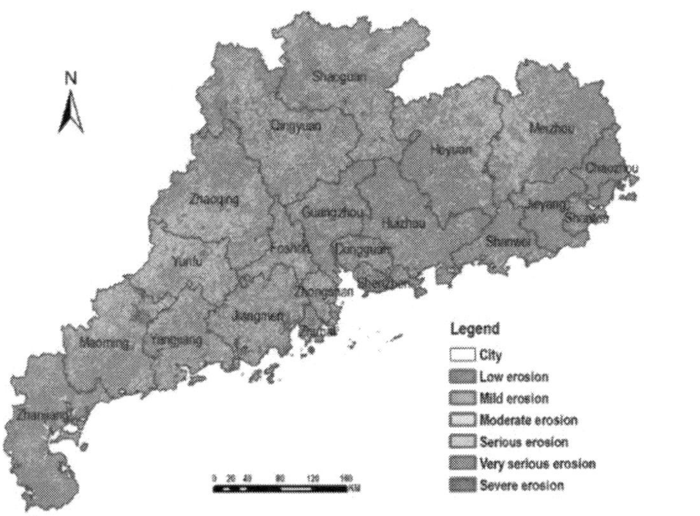


**Figure 1.** The spatial distribution of average annual soil loss in Guangdong in 2012.

As can be seen from the erosion amount percentage, the contribution of the severe erosion is the most, occupied 39.66% of the average annual total erosion; the second contribution of soil erosion level is very serious erosion, with an area of 18.82%; serious erosion, very serious erosion and severe erosion with an area of 11.26% of the total area, but the contribution to the erosion amount reaching 71.66% of the total erosion amount, which is the key area of the soil and water conservation in the future.

According to the analysis of the spatial distribution of soil erosion map in Guangdong in 2012, four high sensitive area of soil erosion was found. These areas include the southern part of Lianzhou City in the northern mountainous region, the junction of Liannan County and Yangshan County, the northern part of Qingxin County, the northern and western parts of Yingde City; West of the cloud fog mountain area of Xinyi, northeast of Gaozhou City, northwest and southwest of Yangchun City; Deqing County in the central west, the northern part of Yunfu City and Gaoyao City; southern Xinhui City and the junction of Taishan City, the southern region of Zhongshan City.

The spatial difference of soil erosion was obvious. The high sensitive area of soil erosion in west is mainly caused by land use/cover types. And slope land is the main land type in the western mountainous region. The north of the study area is the typical Karst area, which was easy to cause soil erosion under heavy rainfall. The southeast of Jiangmen and the southern of Zhongshan were contributed to the rapid development of economy of Pearl River Delta and impact of human activities.

**Table 1**. Various statistic indictors of different soil erosion intensity in 2012

| Intensity | Area/ $km^2$ | Percentage/ % | Average annual erosion module/ ($t.km^{-2}.a^{-1}$) | Average annual total erosion/ ($10^4t.a^{-1}$) | Erosion amount/ % |
|---|---|---|---|---|---|
| Low erosion | 101430.00 | 57.62 | 90.04 | 913.32 | 2.26 |
| Mild erosion | 38919.25 | 22.11 | 1250.10 | 4865.30 | 12.04 |
| Moderate erosion | 15868.25 | 9.01 | 3571.89 | 5667.96 | 14.03 |
| Serious erosion | 8439.25 | 4.79 | 6311.45 | 5326.39 | 13.19 |
| Very serious erosion | 7086.75 | 4.03 | 10727.20 | 7602.10 | 18.82 |
| Severe erosion | 4303.25 | 2.44 | 37223.80 | 16018.33 | 39.66 |
| Total | 176046.75 | 100.00 | 2294.47 | 40393.40 | 100.00 |

## 4. Conclusions

In supports of GIS and MODIS data, the USLE parameters were calculated and the spatial distribution of average annual soil loss of Guangdong was mapped. The average annual soil erosion in Guangdong in 2012 was estimated about 2294.47t/ ($km^2$.a). The soil erosion intensity in the study area mainly focuses on low erosion and mild erosion areas, with an area of 57.62% and 22.11%, respectively. The area of serious erosion, very serious erosion and severe erosion just occupied about 11.26% of the total study area, but the contribution to the average annual soil loss reaching 71.66% of the total erosion amounts. In addition, four high sensitive area of soil erosion in Guangdong for the year 2012 was identified. In general, USLE integrating MODIS data provide an effective way to assess soil erosion quantitatively and efficiently.

## Acknowledgements

China NSFC (41401485) and 863 Program (2006AA06A306) are acknowledged for financial supports.

## References

[1] Zhu M Y 2015 Soil erosion assessment using USLE in the GIS environment: a case study in the Danjiangkou Reservoir Region China *Environ. Earth Sci.* **73** 7899-908
[2] Patil R J, Sharma S K and Tignath S 2015 Remote Sensing and GIS based soil erosion assessment from an agricultural watershed *Arab J. Geosci.* **8** 6967-84
[3] Zhang W B, Xie Y and Liu B Y 2002 Rainfall Erosivity Estimation Using Daily Rainfall Amounts *Scientia Geographica Sinica*, **22** 705-11 (In Chinese with English Abstract)
[4] Sharply A N and Williams J R 1990 EPIC-Erosion/Productivity impact calculator 1: Model Documentation *U.S. Department of Agriculture Technical Bulletin*
[5] McCool D K, Foster G R, Mutchler C K and Meyer LD 1987 Revised slope steepness factor for the universal soil loss equation *Trans. ASAE* **30** 1387–96.
[6] Liu B Y, Nearing M A and Risse L M 1994 Slope gradient effects on soil loss for steep slopes. *Trans. ASAE* **37** 1835–40.
[7] Ma C F, Ma J W and Buhe A 2001 Quantitative assessment of vegetation coverage factor in USLE model using remote sensing data *Bulletin of Soil and Water Conservation* **21** 6–9 (In Chinese with English Abstract)
[8] Huang J L, Hong H S, Zhang L P and Du P F 2004 Study on predicting soil erosion in Jiulong River watershed based on GIS and USLE *J. Soil Water Conserv.* **18** 75–9

IOP Conf. Series: Earth and Environmental Science **74** (2017) 012008 doi:10.1088/1755-1315/74/1/012008

# Tempo-spatial changes of vegetation coverage using remote sensing in Altay, China

**X Y Chen[1, 2], B H Fu[1*] and P L Shi[1]**

[1]Institute of Remote Sensing and Digital Earth (RADI), Kashgar Research Centre, Chinese Academy of Sciences (CAS), Beijing 100094, P. R. China;
[2]University of Chinese Academy of Sciences (UCAS), Beijing 100094, P. R. China

E-mail: chenxiuyan1992@outlook.com

**Abstract.** Vegetation dynamics are very sensitive to climate change, especially in arid and semi-arid regions. In this paper, the Moderate Resolution Imaging Spectroradiometer (MODIS) data were used to evaluate vegetation coverage area in Altay Prefecture during the 2000 to 2015. Furthermore, the patterns of tempo-spatial variations in vegetation coverage were analysed. The vegetation coverage area significantly changed in the past 16 years and the average percentage of vegetation was 35.72%. The results also indicated that the trend of vegetation cover had obvious discrepancy in various altitude ranges.

## 1. Introduction

The Altay Mountain, located on the northern Altay Prefecture, is a trans-country mountain range among China, Russia, Mongolia and Kazakhstan in Central Asia. It is the source of rivers including the Irtysh and Ulungur River, these rivers play an extremely important role in the ecological environment of Central Asia. However, the ecosystem of Altay Prefecture is fragile because of the semi-arid climate. Monitoring and assessing the environment changes in this region is crucial to understand the global change, ecological security and sustainable development.

Vegetation is one of the primary indicators for ecosystem. Changes of vegetation coverage can reflect dynamic variations of regional energy balance, biochemistry cycle, atmospheric cycle and water cycle [1,2]. Meanwhile, it can also indicate the positive and negative impacts of human activities on ecological environment [3,4]. Therefore, aside from evaluating the inter-annual changes of vegetation coverage, we should also monitor the spatial characteristics of changing areas.

[*] Corresponding author.

Considering the broad spatial coverage and long time span in this study, we used MODIS data, which have high spatial and spectral resolutions, to extract vegetation coverage area. In the field of remote sensing, vegetation indexes have been widely used for monitoring vegetation changes [5,6]. EVI (enhanced vegetation index) is more sensitive in high biomass areas and more stable under high aerosols compared to NDVI (normalized difference vegetation index) [6,7]. Therefore, this paper utilized EVI to evaluate vegetation coverage area. Furthermore, Digital Elevation Model (DEM) data from Shuttle Radar Topography Mission (SRTM) were used to analyse the spatial characteristics of vegetation change.

## 2. Study area and data source

### *2.1. Study area*

The study area is Altay Prefecture, which is located on northern Xinjiang Province. It has a geographic coverage of 45°00′- 49°11′N, 85°31′- 91°04′E (figure 1). The west and north of Altay Prefecture is mountainous area, the middle part is valley and the south-eastern part is desert. Altay Prefecture has a cold temperate continental climate with less precipitation and great range of mean annual temperature. Its average temperature is 4.5℃, and annual precipitation is 256.6 mm. Altay Mountain lies in the northern Altay Prefecture, it is nearly 500 km long and 1000-3000 m high. In Altay Mountain, the monthly mean temperature ranges from 24℃ to 40℃ in July and -14℃ to -60℃ in January.

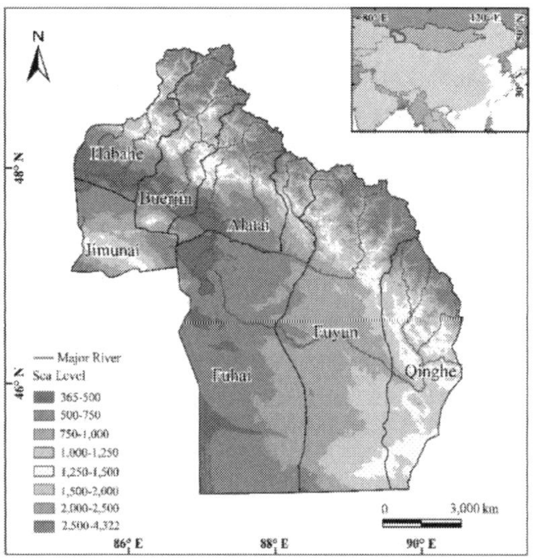


Figure 1. Location map of study area.

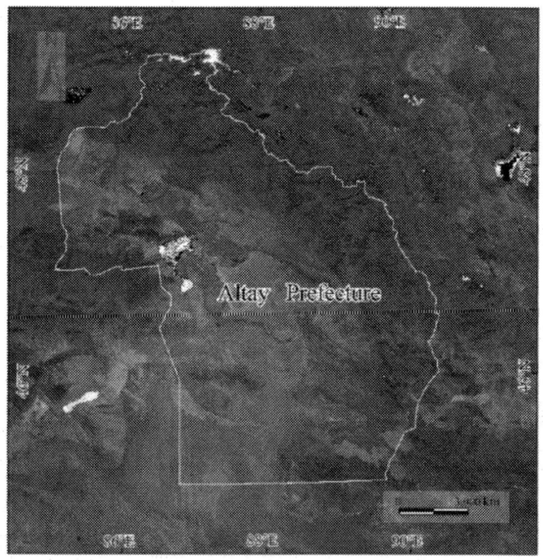


**Figure 2.** MODIS image of study area.

### *2.2. Data source*

The data used in this paper are MODIS MOD13Q1 images with tracks of h23v4 and h24v4. MOD13Q1 data, which were acquired from the Land Processes Distributed Active Archive Center (LP DAAC), are MODIS level 3 product. These data are provided every 16 days at 250 m spatial resolution and in the Integerized Sinusoidal (ISIN) projection. This paper utilized MOD13Q1 EVI bands to extract vegetation coverage area and analyse temporal and spatial changes.

EVI is calculated using surface reflectance values from the red, NIR (near-infrared) and blue bands:

$$EVI = G \times \frac{\rho_{NIR} - \rho_{red}}{\rho_{NIR} + C_1 \times \rho_{red} - C_2 \times \rho_{blue} + L} \quad (1)$$

where $\rho_{NIR}$, $\rho_{red}$ and $\rho_{blue}$ are MODIS surface reflectance of the NIR, red and blue bands, L is the background (soil) adjustment coefficient, and $C_1$, $C_2$ are the fitting coefficients. This equation utilizes the blue band to correct aerosol influence in red band. In the EVI equation, values of these coefficients are, L=1, $C_1$=6, $C_2$=7.5, and G (gain) =2.5 [7].

## 3. Results and Discussion

### *3.1. Temporal changes of vegetation area*

The main contents of pre-processing MODIS EVI data were mosaicking every two images and reprojecting mosaicked images to UTM WGS84 coordinate system. Then we chose 1500 as the optimised EVI threshold to extract vegetation coverage area (table 1). Vegetation area in Altay Prefecture changed significantly during 2000-2015, the average vegetation coverage area was 41,999.87 km$^2$, and average percentage was 35.72%.

**Table 1.** Vegetation area in Altay Prefecture

| year | Vegetation area (km$^2$) | Percentage (%) | year | Vegetation area (km$^2$) | Percentage (%) | year | Vegetation area (km$^2$) | Percentage (%) |
|---|---|---|---|---|---|---|---|---|
| 2000 | 38,224.964 | 32.512 | 2006 | 41,979.343 | 35.705 | 2012 | 42,463.273 | 36.116 |
| 2001 | 47,022.201 | 39.994 | 2007 | 41,968.468 | 35.696 | 2013 | 46,559.333 | 39.600 |
| 2002 | 39,846.876 | 33.891 | 2008 | 37,555.412 | 31.942 | 2014 | 40,168.058 | 34.164 |
| 2003 | 43,338.634 | 36.861 | 2009 | 38,123.653 | 32.425 | 2015 | 40,143.621 | 34.143 |
| 2004 | 42,117.778 | 35.823 | 2010 | 43,681.754 | 37.153 | | | |
| 2005 | 45,148.418 | 38.400 | 2011 | 41,789.033 | 35.543 | | | |

### *3.2. Spatial changes of vegetation area*

There are a variety of geomorphological types in Altay Prefecture, which lead to evident altitude discrepancy in different regions. The great discrepancy in elevation results in an obvious characteristic of vertical distribution in hydrothermal conditions and human activities, and these will potentially influence the vegetation growth [8]. This paper divided the study region into three districts according to altitude: high altitude regions (above 1500 m), middle altitude regions (640-1500 m) and low altitude regions (below 640 m) and extracted vegetation area respectively (figure 3). Figure 3(a) shows: there are more than half of vegetation distribute in high altitude regions. These districts are covered with ice and snow all year round, and thus the vegetation can get adequate meltwater. Figure 3(b) shows: the inter-annual changing trend in these three districts have great differences, there are conspicuous changes in middle and low altitude regions, while small changes in high altitude regions. These results implied that the changes of vegetation coverage area mostly take place in middle altitude regions, low altitude regions will be the next, but rarely in high altitude regions.

In order to validate the above results, the vegetation images of 2001 and 2013 with the largest vegetation area are used to subtract the vegetation image of 2008 with the smallest vegetation area, then we can get figures 4 and 5. In these figures, green pixels stand for increased vegetation area and red pixels are decreased vegetation area. The main vegetation changes from 2001 to 2008 are reduction of area and they mostly distributed in the low altitude regions. And the major changes in vegetation from2008 to 2013 are area increasing and they mainly distributed in the low to middle

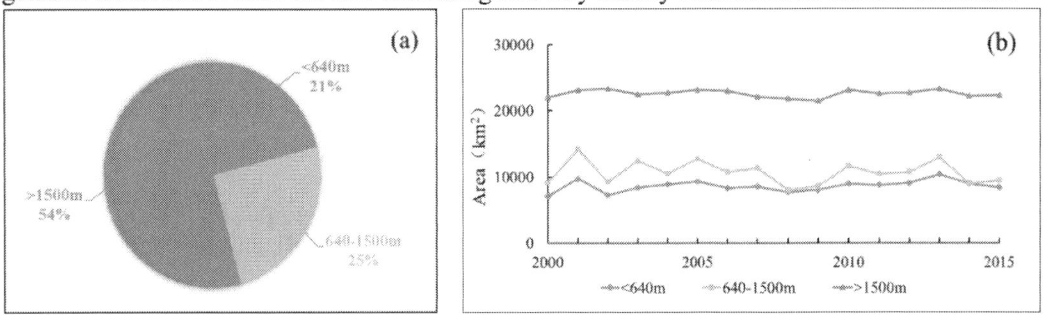


**Figure 3.** Vegetation area percentages(a) and changes(b) in three districts.

altitude regions. The amount of vegetation area in the high altitude regions are almost stable from 2000 to 2015 as shown in figure 3(b).

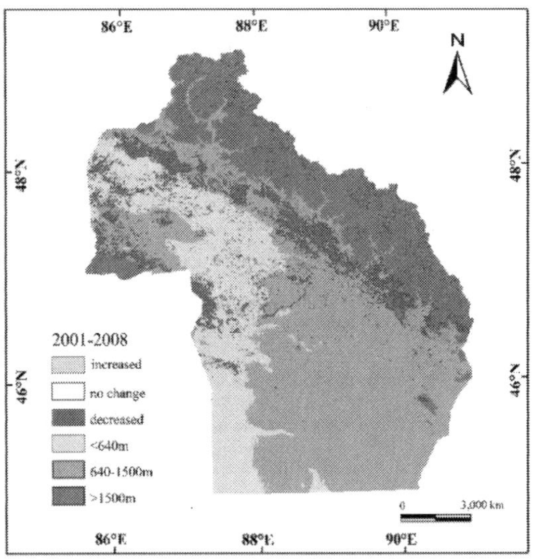


**Figure 4.** Vegetation changes from 2001 to 2008.

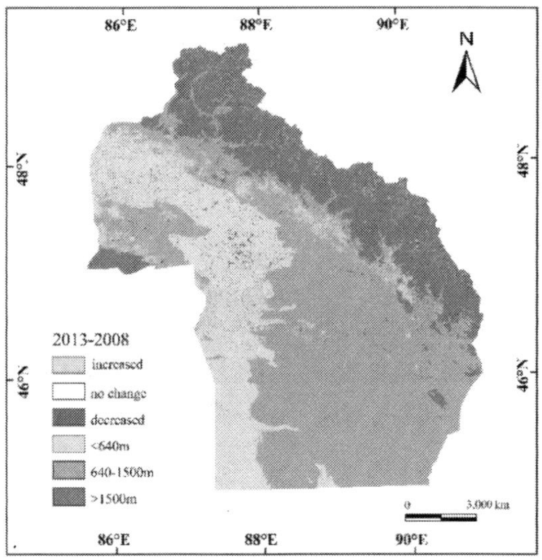


**Figure 5.** Vegetation changes from 2008 to 2013

### *3.3. Climate control factors on vegetation dynamics*

Previous study showed that both the temperature and precipitation have increased significantly (0.35 ℃ /10y, 11.2 mm/10y) in northern Xinjiang during the period 1960-2011 under global warming [9]. As a sensitive indicator of ecosystem, the vegetation cover has changed. Many studies have documented that both the temperature and precipitation are major control factors on the growth of arid and semi-arid vegetation in northwestern China [1], but the precipitation is the dominant one [10,11].

## 4. Conclusions

In this study, the tempo-spatial features of vegetation change in Altay Prefecture from 2000-2015 are extracted using the MODIS EVI images, and we got the following conclusions:

(1) There are extremely significant variations of vegetation coverage in the Altay Prefecture during 2000-2015.

(2) The percentages of vegetation distribution in high, middle and low altitude regions are 54%, 25%, 21%, respectively.
(3) The inter-annual changes of vegetation coverage area had great discrepancies during different altitude regions.

**References**

[1] Zhang C, Lu D S, Chen X, Zhang Y M, Maisupova B and Tao Y 2016 The spatiotemporal patterns of vegetation coverage and biomass of the temperate deserts in Central Asia and their relationships with climate controls *Remote Sens. Environ.* **175** 271-81

[2] Kefi S, Rietkerk M, Alados C L, Pueyo Y, Papanastasis V P, Elaich A and De R P C 2007 Spatial vegetation patterns and imminent desertification in Mediterranean arid ecosystems *Nature* **449** 213-7

[3] Wang J, Wang K L, Zhang M Y and Zhang C H 2015 Impacts of climate change and human activities on vegetation cover in hilly southern China *Ecol. Eng.* **81** 451-61

[4] Eisenfelder C, Kuenzer C and Dech S 2012 Derivation of biomass information for semi-arid areas using remote sensing data *Int. J. Remote Sens.* **33** 2937-84

[5] Huete A, Didan K, Miura T, Rodriguez E P, Gao X and Ferreira L G 2002 Overview of the radiometric and biophysical performance of the MODIS vegetation indices *Remote Sens. Environ.* **83** 195-213

[6] Huete A, Justice C and Liu H 1994 Development of vegetation and soil indices for MODIS-EOS *Remote Sens Environ* **49** 224-34

[7] Setiawan Y, Yoshino K and Prasetyo L B 2014 Characterizing the dynamics change of vegetation cover on tropical forestlands using 250 m multi-temporal MODIS EVI *Int. J. Appl. Earth Obs. Geoinformation* **26** 132-44

[8] Zapata R X, Brooks P D, Troch P A, McIntosh J and Guo Q H 2016 Influence of terrain aspect on water partitioning, vegetation structure and vegetation greening in high-elevation catchments in northern New Mexico *Ecohydrology* **9** 782-95

[9] Xu C C, Li J X, Zhao J, Gao S T and Chen Y P 2015 Climate variations in northern Xinjiang of China over the past 50 years under global warming *Quat. Int.* **358** 83-92

[10] Jin X M, Liu J T, Wang S T and Xia W 2016 Vegetation dynamics and their response to groundwater and climate variables in Qaidam Basin, China *Int. J. Remote Sens.* **37** 710-28

[11] Jiapaer G, Liang S L, Yi Q X and Liu J P 2015 Vegetation dynamics and responses to recent climate change in Xinjiang using leaf area index as an indicator *Ecol. Indic.* **58** 64-76

IOP Conf. Series: Earth and Environmental Science **74** (2017) 012009 doi:10.1088/1755-1315/74/1/012009

# Simulation and Evaluation of Low Impact Development of Urban Residential District Based on SWMM and GIS

**Tielan Huang[1], Yunpeng Wang[2*], Jinlan Zhang[1]**
1 Guangdong Polytechnic of Industry and Commerce, Guangzhou 510510, China
2 Guangzhou Institute of Geochemistry, Chinese Academy of Sciences, Guangzhou 510640, China

E-mail: wangyp@gig.ac.cn

**Abstract.** In this study, simulation and evaluation of low impact development in resident district was carried out based on Storm Water Management Model (SWMM) and GIS method. In our model, we added 3 kinds of low impact development facilities, namely permeable pavement, rainwater garden, and green roof. These facilities are used alone or in combination. The model was run under five different rainfall reappearing periods. The simulation results using low impact development facilities were compared with simulation results under the current situation and undeveloped state. The results show that the total amount of runoff was greatly reduced by using various types of low impact development facilities in the urban residential district. The maximum reduction rate was using permeable pavement, reached 29.9%, followed was using rainwater garden, and the worst was using green roof. The lowest cost of reduction of the total amount of runoff was using permeable pavement, the followed was using rainwater garden, and the highest was using green roof. The combination scheme of various low impact development facilities has the highest efficiency of reducing total amount of runoff, and the lowest cost, which considering of the actual situation of the study area. The study indicated that application of low impact development facilities can reduce surface runoff effectively, which should be a useful way for prevention of urban waterlogging.

## 1. Introduction

In recent years, with the rapid development of urbanization in China, city impervious areas were increased sharply, and urban water environment was greatly changed. Urban waterlogging has become a serious environmental disaster. Especially in the residential district of the city, the urban waterlogging resulting in ground flooding, traffic jams, underground submerging, and interruption of utilities pipeline such as gas, electricity. Urban waterlogging seriously impacted on people's life and property safety. In the past, the control schemes of urban waterlogging were focused on drainage. Those schemes can achieve a certain effect in the short term, but cannot solve the problem fundamentally.

Low impact development model (LID) can simulate the natural hydrological conditions as much as possible. It emphasized the combination of small scale, sustainable, low energy consumption, and the combination of landscape design and land development [1]. It can effectively repair hydrological mechanism which was destroyed by the urban drainage system [2-3]. Therefore, LID has been widely used in the control and utilization of rain flood in developed countries in recent years [4-6].

* Corresponding author.

IOP Conf. Series: Earth and Environmental Science **74** (2017) 012009 doi:10.1088/1755-1315/74/1/012009

The SWMM model was developed by the national environmental protection agency. It was mainly used to simulate the rainfall-runoff process of a single precipitation event or a long series of rainfall series. SWMM included hydrological, hydraulic and water quality modules. LID module was increased in the latest version of SWMM5.0 [7-8]. It can simulate 5 kinds of low impact development measures, such as bioretention, infiltration ditch, permeable pavement, rain barrels, and vegetation swales. Other LID measures, such as filter belt, down concave type green space, green roof, can be simulated by the parameters transformation.

Therefore, we simulated low impact development in resident district by SWMM and GIS.

## 2. Meterials and Method

*2.1 Study area and data source*

The study area, Zhonghaikangcheng District, is located at Tianhe District of Guangzhou City, China. The elevation ranges is from 2.66 to 10.05 m above mean sea level, higher at the northeast, and lower at southwest. It covers the area of approximately 25,400 $m^2$. The study area enjoys a humid subtropical climate with abundant rainfall. The annual average temperature is between 21 and 24℃, and the annual mean precipitation reaches 1,721.5 mm. The annual rainfall is concentrated in June to September. Winter is short, mild and relatively dry, while summers are long, hot and very wet.

The data used in the study area mainly include the QuickBird 0.61 m remote sensing image at 2013 (used to extract the land use type), the topographic map of Guangzhou with a scale of 1:10,000 (used to build DEM, and to extract elevation and slope), construction planning map of the study area with a scale of 1:10,000(used to adjust sub catchments).

*2.2Method*

We simulated of the low impact development of urban residential areas based on SWMM and GIS. The main steps included: generalization of the study area, design of heavy rain process line, setting of LID scenes.

*2.2.1 Generalization of the study area*

First, we extracted landuse/landcover types of the study area. They data sources are topographic map of Guangzhou with a scale of 10,000, QuickBird 0.61 m remote sensing image at 2013, and data of field investigation. There were 4 landuse types: greenbelt, water area, roof, and road (Figure 1). We used graphic editing and spatial analysis functions of ArcGIS 10.2 to carry out the task.

Second, we extracted DEM from topographic map of Guangzhou. Then we divided sub catchments based on DEM, and adjusted the boundary of sub catchments according to road traffic map. The study area is divided into 38 sub catchments, and each sub catchment area has a water node. The parameters of sub catchments can be obtained by spatial analysis fuctions ArcGIS 10.2.

Finally, we built drainage network of the study area. Drainage pipes are mainly distributed at the boundary of each sub catchment. Some pipes must be adjusted according to terrain of corresponding sub catchment. 77 drainage pipes and 84 pipeline nodes were built, including 83 inspection wells, 1 export nodes. The parameters of drainage pipes and nodes can also be obtained by ArcGIS 10.2.

The Generalized Map of study area is as Figure 1.

*2.2.2 Design of heavy rain process line*

The study area is located in Guangzhou City, and the formula of rainstorm intensity is as follow:

$$q = \frac{2424.17(1 + 0.553 \lg T)}{(t + 11.0)^{0.668}}$$

Where q is heavy rain intensity (L/s • ha), T is the reappearing period (a), and t is the rainfall duration (min).

We designed the heavy rain process line of the study area based on formula of rainstorm intensity of Guangzhou City and Chicago rain type. The reappearing periods are 1, 5, 10, 50 and 100 years. The interval is 5 minutes.

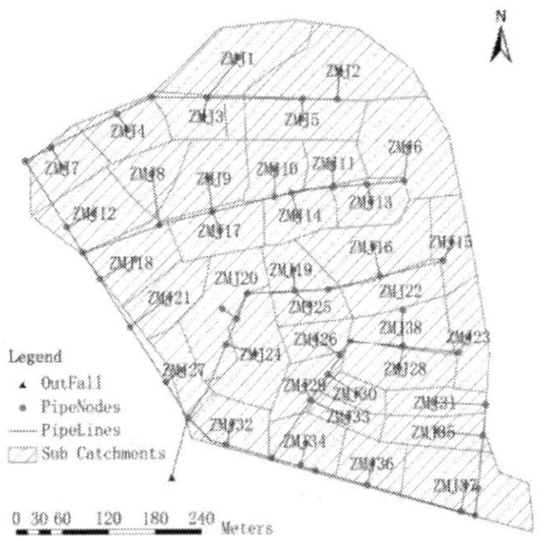


Figure.1 Generalized Map of the study area

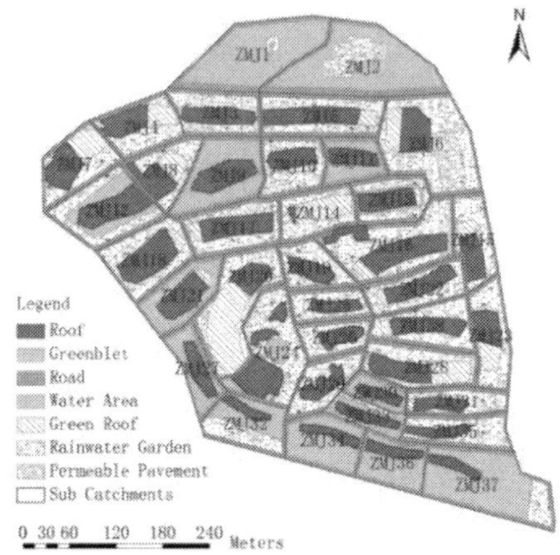


Figure.2 Combination Scheme of LIDS

*2.2.3 Setting of LID scene*

After analysis of the actual situation of the study area, we selected 3 kinds of LID facilities, such as green roof, permeable pavement, and rainwater garden. There were 5 scenarios: 1) Undeveloped state, all the ground of the study area is green; 2) Current status, the study area was developed; 3) All greenbelt were transferred to rainwater garden, do not add any other LID facilities; 4) All road and were transferred to permeable pavement, do not add any other LID facilities; 5) All roof were transferred to green roof, do not add any other LID facilities. 6) Combination scheme of 3 kinds of LID, according to comparison of the efficiency of cutting the amount of total runoff, cost, and the actual situation of each sub catchment area.

The combination scheme (Figure.2) is: area of green roof is 14056.2m$^2$(13.3% of total roof), area of permeable pavement is 57431.1 m$^2$ (64.9% of total road), area of rainwater garden is 52921.9m$^2$(78.1% of total greenbelt).

## 3. Result and Discussion

Under different reappearing periods, the simulation results of total runoff and reduction percentage as shown in Table 1, by using different LID scenarios. The comparison of cost of reduction under different scenarios is shown in Table 2.

Table 1 Runoff simulation results under different reappearing periods
(Scenarios: 1. Undeveloped; 2. Current status; 3. Using green roof separately; 4. Using permeable pavement separately; 5. Using rain-water garden separately; 6. Combination of 3 kinds of LID facilities)

| Scenarios | Total runoff(m$^3$) | | | | | Reduction percent of total runoff (%) | | | | |
|---|---|---|---|---|---|---|---|---|---|---|
| | P=1 | P=5 | P=10 | P=50 | P=100 | P=1 | P=5 | P=10 | P=50 | P=100 |
| 1 | 5415.6 | 10867.8 | 13,317.60 | 19114.2 | 21645 | | | | | |
| 2 | 12826.8 | 18859.8 | 21473.4 | 27556.8 | 30182.4 | | | | | |
| 3 | 10920.6 | 16963.8 | 19584.6 | 25667.4 | 28294.8 | 14.5 | 10.1 | 8.5 | 6.9 | 6.3 |
| 4 | 8496.0 | 13012.8 | 15040.8 | 19801.2 | 21921.6 | 33.5 | 31.0 | 29.7 | 28.1 | 27.4 |
| 5 | 9323.4 | 14524.8 | 16875.6 | 22561.8 | 25077.6 | 27.0 | 23.0 | 21.1 | 18.1 | 16.9 |
| 6 | 5402.4 | 9672.6 | 15040.8 | 16570.8 | 18840.6 | 57.9 | 48.7 | 30.0 | 39.9 | 37.6 |

Table 2 Comparison of cost of reduced total runoff under different scenarios

| | Using green roof separately | Using permeable pavement separately | Using rainwater garden separately | Combination of 3 kinds of LID facilities |
|---|---|---|---|---|

| | | | | |
|---|---|---|---|---|
| Area（$m^2$） | 80851.8 | 87311.2 | 85241.8 | 124409.2 |
| Entire cost($10^4$ Yuan) | 2162.8 | 1309.7 | 1022.9 | 1859.0 |
| Reduced total runoff($m^3$) | 1893.6 | 6525.7 | 4507.2 | 13105.4 |
| Cost of reduced total runoff(Yuan/$m^3$) | 11421.6 | 2007.0 | 2269.5 | 1418.5 |

As seen from the tables, compared with the current situation (after development), the total runoff were reduced by using any LID facilities separately, or under combination sense. Under different scenarios, with the increase of the reappearing period of rainfall, the reduced proportion of total runoff decreased gradually. Reduction rate of total runoff by using permeable pavement is highest (29.9%), the second is using rainwater garden (21.2%), and the lowest is using green roof (9.7%). Cost of reduced total runoff by using permeable pavement is lowest (2007.0Yuan/$m^3$), the second is using rainwater garden (2269.5 Yuan/$m^3$), and the highest is using green roof (11421.6 Yuan/$m^3$). Compared with the application of one LID separately, reduction effect of total runoff is greatly improved by combination of 3kinds LID facilities. Average reduction rate reached 42.8%. Cost of reduced total runoff is decreased significantly, is 1418.5 Yuan/$m^3$. The cost of all program by using LID facilities is too high, more than 10 million Yuan.

## 4. Conclusions

Through this study, conclusions are as follows:
(1) Total runoff was greatly reduced by using one LID facilities separately, or under combination scheme of 3 kinds of LID facilities. Therefore, we should vigorously promote the LID concept and measures in the construction of new urban areas and the transformation of the old city.
(2) In practical work, we should consider the environment of urban residential district, cost of different LID facilities, and the drainage capacity of municipal pipe network. The optimal LID layout scheme must achieve the overall optimization of economic and social benefits.

## Acknowledgements

China NSFC (41401485) and 863 Program (2006AA06A306) are acknowledged for financial supports.

## References

[1] He S, Liu J and Zhu J Q 2013 Rainwater Control and Utilization Effect Assessment and Simulation of Low Impact Development Based on SWMM *International Journal Hydroelectric Energy* **31** 42-5 (In Chinese with English Abstract)
[2] Liu H 2014 Development Strategies of Urban Waterfront Space Based on the Concept of Low Impact Development *Journal of Guangdong Landscape Architecture* **36** 29-32 (In Chinese)
[3] Wang H W, Mao Y F, Gao Y, et al. 2012 Low Impact Development Practice and Its Effects *Environ. Sci. Technol.* **35** 99-103
[4] Bedan E S, Clausen J C 2009 Stormwater Runoff Quality and Quantity From Traditional and Low Impact Development Watersheds *Jawra Journal of the American Water Resources Association* **45** 998-1008
[5] Dietz M E 2007 Low Impact Development Practices: A Review of Current Research and Recommendations for Future Directions *Water, Air, Soil Pollut.* **186** 351-63
[6] Hunt W F, Traver R G, Davis A P, et al. 2010 Low Impact Development Practices: Designing to Infiltrate in Urban Environments *Effects of Urbanization on Groundwater*
[7] Zhao D Q, Tong Q Y, Wang H Z, et al. 2009 Application of SWMM Model in Analysis of Urban Rainwater Drainage System *China Water & Wastewater,* **5** 198-201 (In Chinese)
[8] Wang K, Gao C, Zhu J Q, et al 2014 Compensation Mechanism of Permeable Channel LID Measures Based on SWMM *International Journal Hydroelectric Energy* **32** 19-21

IOP Conf. Series: Earth and Environmental Science 74 (2017) 012010 doi:10.1088/1755-1315/74/1/012010

# Monitoring an air pollution episode in Shenzhen by combining MODIS satellite images and the HYSPLIT model

**Lili Li[1], Yihong Liu[2], Yunpeng Wang[1*]**

1 Guangzhou Institute of Geochemistry, Chinese Academy of Sciences, Guangzhou 510640, China

2 School of Geographic and Oceanographic Sciences, Nanjing University, Xianlin Road No.163, Qixia District, Nanjing 210046, China

E-mail:wangyp@gig.ac.cn

**Abstract.** Urban air pollution is influenced not only by local emission sources including industry and vehicles, but also greatly by regional atmospheric pollutant transportation from the surrounding areas, especially in developed city clusters, like the Pearl River Delta (PRD). Taking an air pollution episode in Shenzhen as an example, this paper investigates the occurrence and evolution of the pollution episode and identifies the transport pathways of air pollutants in Shenzhen by combining MODIS satellite images and HYSPLIT back trajectory analysis. Results show that this pollution episode is mainly caused by the local emission of pollutants in PRD and oceanic air masses under specific weather conditions.

## 1. Introduction

As a coastal city located in south-central Guangdong, China (about 113.7°E -114.6 and 22.4 °N-22.9 °N), Shenzhen is one of areas that boasts fast economic growth and population explosion in the world. Rapid industrial development and heavy traffic pressure caused the increase in haze days and deterioration of air quality, which has drawn widespread attention from the government and public[1-2]. However, air pollution is not only a local but a regional problem, for some pollutants could be transported over hundreds and thousands of kilometres [3]. Owing to its unique topography and meteorology, Shenzhen is sensitive to regional transport of air pollutants from the surrounding areas.

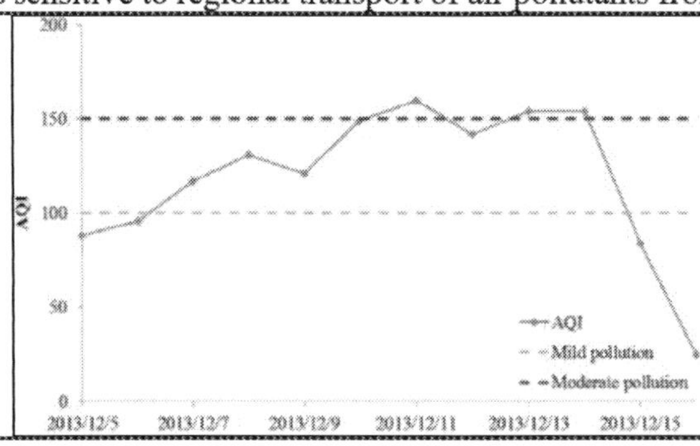


**Figure 1.** Variation trend of air quality index(AQI) during December 5-16, 2013 in Shenzhen

1 Corresponding author.

During December 5th to 16th, 2013, an air pollution episode occurred in Shenzhen, and the air quality index (AQI) reached the standard of mild pollution from December 7, and then increased into moderate pollution during December 10-14, as is shown in Figure 1.Taking this episode as an example, this study combines MODIS satellite images with the Hybrid Single-Particle Lagrangian Integrated Trajectory (HYSPLIT) model to analyse the pollution episode and identify the transport pathways of air pollutants in Shenzhen.

## 2. Data and Method

### 2.1. Data

Satellite data used in this study over Shenzhen and its surrounding areas were provided by a Moderate-resolution imaging spectroradiometer (MODIS) (on-board the Terra and Aqua satellite) and downloaded from NASA Level 1 and Atmosphere Archive and Distribution System. They are calibrated L1B data with 500m spatial resolution including visible, near-infrared, and mid-infrared bands. Images during December 6-14, 2013 were acquired, together with monitoring data from 11 air quality stations provided by the Department of Environmental Protection of Guangdong Province, China. The station observations include air quality index, and concentrations of PM10, PM2.5, SO2, NO2 and Ozone.

### 2.2. HYSPLIT model

The Hybrid Single-Particle Lagrangian Integrated Trajectory (HYSPLIT) transport model, developed by NOAA's Air Resources Laboratory, is a complete system for computing simple air mass trajectories, as well as complex transport, dispersion, chemical transformation, and deposition simulations[4]. The model calculation method is a hybrid between the Lagrangian approach, using a moving frame of reference for the advection and diffusion calculations as the trajectories or air parcels move from their initial location, and the Eulerian methodology, which uses a fixed three-dimensional grid as a frame of reference to compute pollutant air concentrations[5]. The model can be run as a stand-alone application, or interactively on the website (http://www.ready.noaa.gov/index.php).

## 3. Results and discussion

According to the variations of AQI in Figure 1, the air pollution episode is divided into four phases: before the pollution (December 5-6), beginning of the pollution (December 7-9), worsening of the pollution (December 10-14) and disappearance of the pollution (December 15-16). Figure 2 is MODIS images covering Shenzhen (the red polygon) and its surrounding areas on Dec. 7, 9, 10 and 13. Images below clearly show the occurrence and transportation of the air pollution episode. In the satellite image on Dec. 6, ground objects were clear with little cloud cover, indicating that air quality over the region was fine. On Dec. 9, the air pollution episode broke out, and the thick cloud and aerosol layer led to the low visibility of the objects. Oceanic air masses from the east-south moved to Shenzhen, and accelerated the air pollution on Dec. 10. Until December 13, the whole region was covered with thick cloud cover and haze so that nothing could be recognized from the MODIS image, and the air quality turned to be moderate pollution.

Taking Shenzhen(22.5°N, 114°E)(UTC+0800) as the reference point, we used HYSPLIT model to calculate the backward air masses trajectories at the altitude of 50m, 100m and 300m in four periods of this air pollution episode, as Figure 3 shows. The red, blue and green lines represent the path of air masses at 50 meters, 100 meters and 300 meters, respectively. The region was affected by different airflows in the pollution period, as was controlled by the north air masses before and after the pollution Shenzhen, and during the pollution process was controlled by the south air masses.

On December 1, a large-scale haze pollution event broke out in the central and eastern parts of China, and the northerly air from the inland areas blew the polluted air to the southern coastal areas. But due to the weak transportation the north wind, the air quality in Shenzhen was fine. But until December 7th, the south-east air turned to be oceanic air mass from the South China Sea Strait in the east, and arrived in Shenzhen along the south China coastline. The near-surface airflow mainly originated from the east Taiwan Strait. The weak south-east wind facilitated the accumulation of surrounding pollutants, thus mild air pollution occurred in Shenzhen. During December 10 to 14, the trajectory of the near-surface air in Shenzhen came from the Pearl River Delta, through the South China Sea and the Pearl River Estuary, acting as a loop; air at low altitudes was influenced by the oceanic air mass from the northeast, and reached Shenzhen through the southeast coast, the western Guangdong and the Pearl River estuary. The trajectory of air was above 2000hpa and controlled by high pressure, thus the stable atmospheric structure of troposphere hindered the diffusion of the pollutants and caused moderate pollution in the region. After December 15, the direction of air changed to north-east, and the air quality in Shenzhen gradually improved.

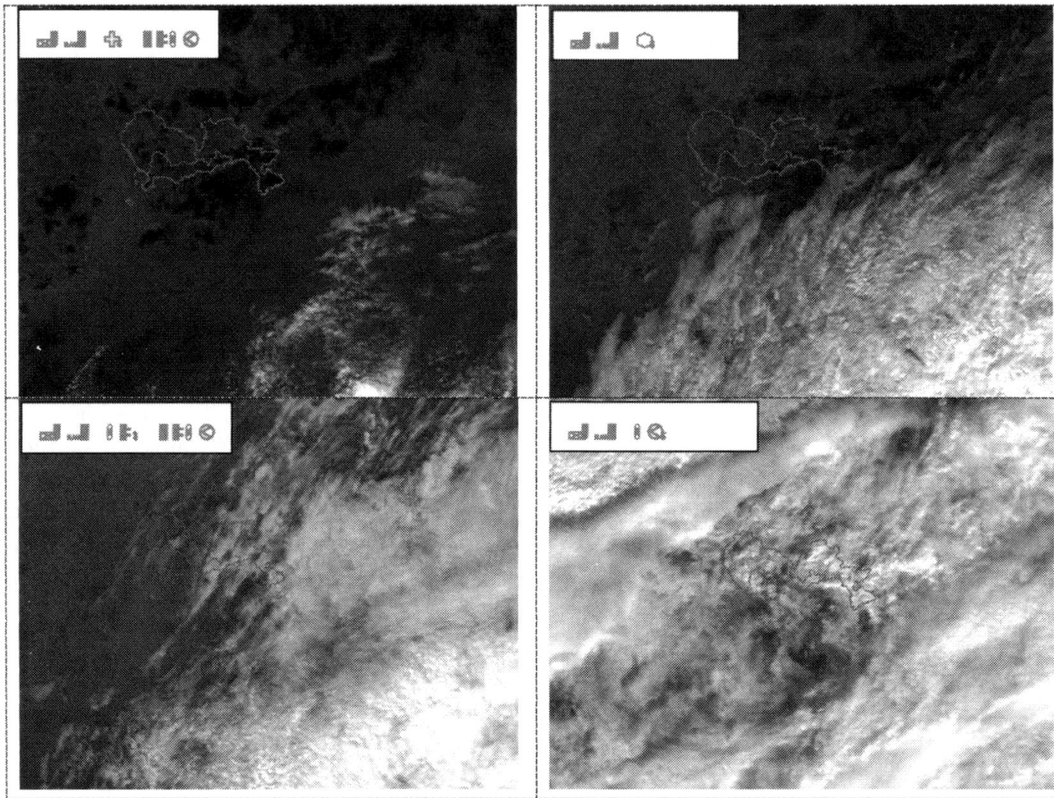

**Figure 2.** MODIS images covering Shenzhen on Dec. 7, 9, 10 and 13, 2013.

## 4. Conclusions

Combining MODIS satellite images and back trajectories from HYSPLIT model, this study analysed the four phases of an air pollution episode during December 5-16, 2013 in Shenzhen and identified the transport pathways of air pollutants. The results indicate that weak east-south wind and weak atmospheric vertical movement were the weather conditions during the pollution, while the north airflow and strong air convection after December 15th were conducive to the diffusion of air pollutants. Backward trajectory analysis shows that the pollution episode was mainly caused by the local emission of pollutants in the Pearl River Delta region and oceanic air masses under specific weather conditions.

## Acknowledgement

China National 863 Program (2006AA06A306) and State Key Laboratory of Organic Geochemistry Program (SKLOGA201603A) are acknowledged for financial supports.

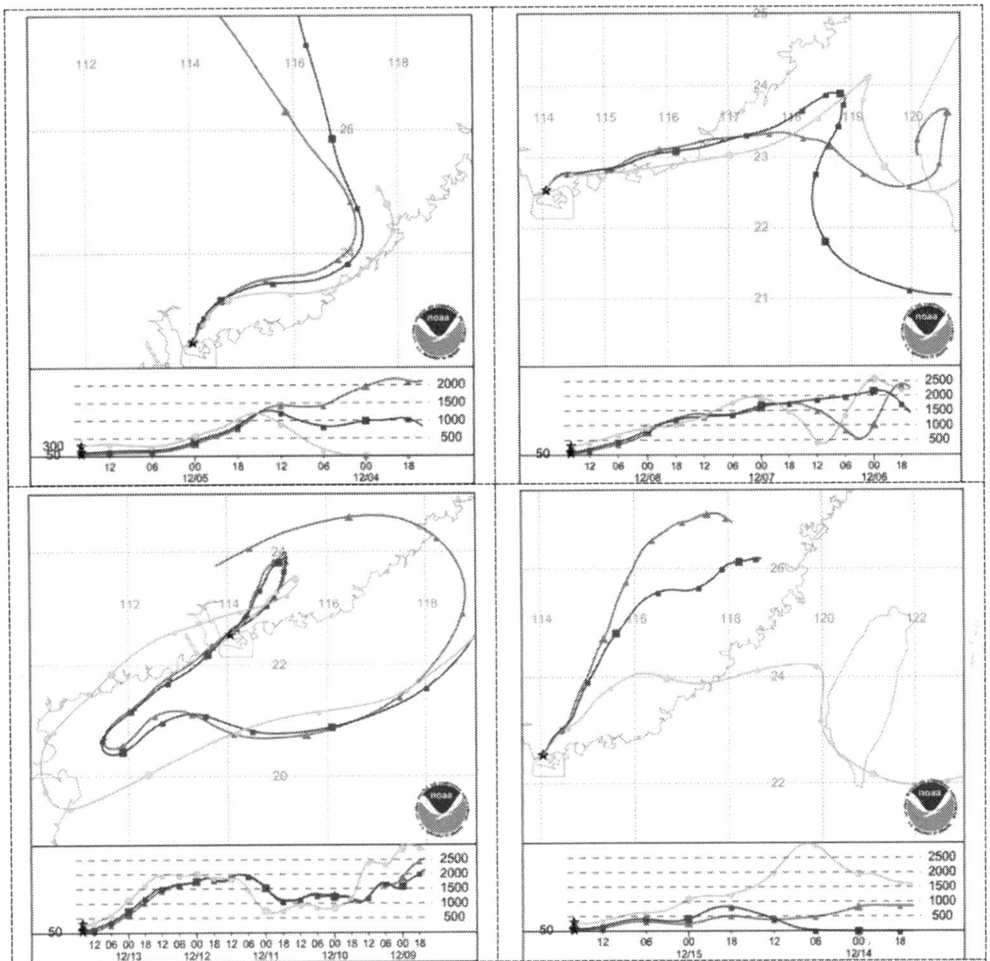

**Figure 3.** The backward air masses trajectories in Shenzhen in four periods( December 5-6, 7-9, 10-14 and 15-16, 2013)

**References**

[1] Deng X J, Tie X X, Wu D, Zhou X J, Bi X Y, Tan H B, Li F, Jiang C L 2008 Long-term trend of visibility and its characterizations in the Pearl River Delta (PRD) region, China *Atmos. Environ.* **42** 1424-35

[2] Xie P, Liu X Y, Liu Z R, Li T T, Zhong L J, Xiang Y R 2011 Human health impact of exposure to airborne particulate matter in Pearl River Delta, China. *Water, Air, Soil Pollut.* **215** 349–63

[3] Wang F, Chen D S, Cheng S Y, Li J B, Li M J and Ren Z H 2010 Identification of regional atmospheric PM 10 transport pathways using HYSPLIT, MM5-CMAQ and synoptic pressure pattern analysis Environ. Modell. Softw. **25** 927-34.

[4] Draxler R R, Rolph G D 2015 HYSPLIT (HYbrid Single-Particle Lagrangian Integrated Trajectory) Model access via NOAA ARL READY Website (http://ready.arl.noaa.gov/HYSPLIT.php). NOAA Air Resources Laboratory, Silver Spring, Maryland

[5] Stein A F, Draxler R R, Rolph G D, Stunder B J B, Cohen M D and Ngan F 2015 NOAA's HYSPLIT atmospheric transport and dispersion modeling system *B. Am. Meteorol. Soc.* 96, 2059-77

# Quantitative analysis of land use and land cover changes from the multi-temporal remote sensing data in the Bosten Lake Basin, Chinese Tian Shan

**A Maimaiti[1,2], L M Wang[1*] and F Yan[1], J Zhang[1], Y X Ma[1]**
[1]Kashgar Research Center of Institute of Remote Sensing and Digital Earth, Chinese Academy of Sciences, Kashgar 844000, China;
[2] Key Laboratory of Oasis Ecology, College of Resources and Environmental Sciences, Xinjiang University, Urumqi, 830046, China)
E- mail: lmwang@ceode.ac.cn

**Abstract:** In this paper, the spatial changes of land use and land cover (LUCC) in Bosten Lake Basin from 1985 to 2015 were analyzed, based on the 3S technology and four periods of remote sensing images in 1985, 1998, 2008 and 2015. The driving force of the LUCC was analyzed quantitatively using principal component analysis (PCA) method. The results showed that there was a wide change of LUCC in the Bosten Lake Basin in the past 30 years. During this periods, natural grassland decreased with a rate of 62.6 $km^2/a$, while cultivated land and residential land increased with a rate of 28.9 $km^2/a$. The rapid expansion of cultivated land was the result of natural grassland and unused land reclamation. Meanwhile, other land use types changed slightly. The results of PCA analysis indicated that the LUCC change was result of the interaction of human activity (social and economic factors) and natural environmental changes (climate change). Therefore, it is necessary to consistently improve the natural environment of Bosten Lake Basin. This study can provide stable basis of the theory and practice for sustainable development of Bosten Lake Basin.

**Key words:** Bosten Lake Basin; Land use land cover; Human activities; Climatic change

## 1. Introduction

The dynamics of regional land use and land cover change (LUCC) play an important role in present strategies and policies for natural resource management and monitoring. In arid and semi-arid

* Corresponding author.

environments, the LUCC often reflects the most sensitive indicator for the environmental changes because of human activities and natural forces[1]. At present, many international big projects, including the International Geosphere-Biosphere Program (IGBP), the International Human Dimension Program (IHDP), the World Climate Research Program (WCRP), and the International Program of Biodiversity Science (IPBS) have begun to research the relationship between land use and the hydrological cycle [2]. Recently, numbers of research results have showed that human being can understand natural ecosystem changes in the certain level by analyzing the modes of land use in the past, present, and future [3]. In addition, LUCC is an important factor that can modify hydrological traversing in the range of temporal and spatial scales in arid and semi-arid area.

## 2 Materials and methods

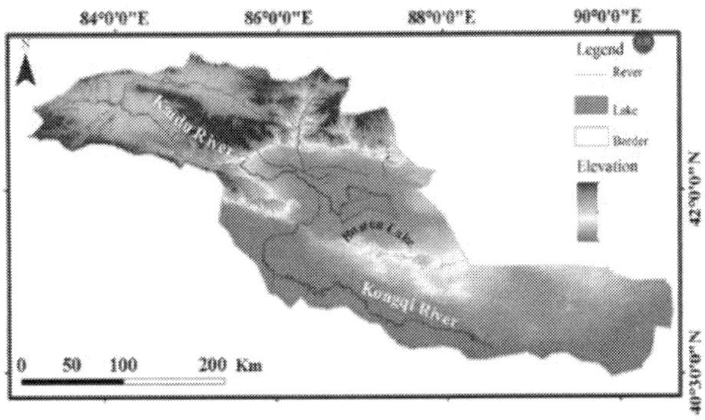


**Figure 1.** Location of the Bosten Lake Basin.

*2.1 Study Area*

Bosten Lake Basin (41°56′ N – 44° 40′ N; 86°14′ E – 90°56′ E) is located on the southern slope of the Tianshan Mountains and lies in the southeastern part of Yanqi Basin (Figure.1). The Yanqi Basin borders the Tianshan Mountains and the Kuruktag Mountains in the northand west, and it linked with Taklimakan desert in the south. Bosten Lake is one of the largest inland freshwater lakes in China with a surface area of 1000 $km^2$. The maximum water depth is 16.2 m, while an average depth is 8 m.

*2.2 Data collection and data processing*

This study used remote sensing images, climate data and Socio-economic statistical data to analyze the LUCC situation in the study area in the past 30 years and to assess the driving force quantitatively. The remote sensing data include Landsat TM / ETM and OLI image in 1985, 1998, 2008, 2015 respectably. In this paper, the data processing has completed using ENVI 5.0 and ArcGIS 10.2 software. We obtained the LUCC data from TM / ETM / OLI imagery, visual interpretation was based on image characteristics such as color, shape, size, shading, texture, structure, and relative spatial distribution of each class of land cover. We used a six-type classification system provided by the Resources and Environment Database of Chinese Academy of Sciences including cultivated land, forestland, grassland, water body, residential and industrial land, and unused land.

*2.3 Study methods*

(1) LUCC dynamic degree model

In this paper, regional difference in land use characteristics were determined using the land use dynamic degree model that could be mathematically expressed by the following relationship [4]:

$$S = \sum_{1}^{n} (\frac{S_n - S_i}{S_i}) \times 100\%$$

where $S$ is the LUCC rate, $S_i$ is the beginning of the monitoring period, $n$ is the time period, and $S_n$ is ending of the monitoring period. In a comprehensive manner, the dynamic degree represented the change of land use in the study region.

(2) Estimate the contribution of driving forces

This paper utilized PCA to assess the contribution rate of human activity and climatic force on LUCC changes in BLB.

## 3. Results and discussion

### *3.1 The characteristics of LUCC in BLB*

The big change of land use and land cover is emerged in the middle part of the BLB. Table 1 and Figure 2 showed that cultivated land, grassland, unused land and water land were the main land use types in BLB in past thirty years, and they accounted for about 98.8% of the total area. Those land use types were responsible for the 92.2% of land use change during whole periods (1985~2015).

**Table 1.** The and use type and their proportional change in BLB.

| Land use type | 1985 ( km$^2$) | 1998 ( km$^2$) | 2008 ( km$^2$) | 2015 ( km$^2$) | 1985-1998 (%) | 1998-2008 (%) | 2008-2015 (%) | 1985-2015 (%) |
|---|---|---|---|---|---|---|---|---|
| cultivated land | 2738 | 2872 | 3262 | 3606 | 4.9 | 13.6 | 10.6 | 31.7 |
| forest land | 754 | 804 | 935 | 827 | 6.6 | 16.3 | -11.6 | 9.7 |
| grassland | 26045 | 25864 | 23935 | 24167 | -7 | -7.5 | 0.97 | -7.2 |
| water area | 6185 | 6129 | 6256 | 6259 | 0.9 | 2.1 | 0.05 | 1.2 |
| residential land | 163 | 395 | 468 | 547 | 34.8 | 18.5 | 16.9 | 86.68 |
| unused land | 38954 | 38775 | 39983 | 39433 | -12.6 | 3.1 | -1.4 | 1.6 |

Land use types including cultivated land, forest land, residential land and unused land increased respectively by 31.7%, 9.7%, 1.2%, 86.68%, 1.6% during the entire period, while the areas of grassland decreased by 7.2% in the same period. Figure. 2 shows that the LUCC of the study area mainly occurred in the southeast part of the Hejing county and middle part of BLB, as well as upper part of Yuli county. It can be known that there has been a wide change in land use and land cover in BLB in recent 30 years. The order of the change from large to small is as follows: residential and industrial land> cultivated land> forest land> grass land> unused land> water land. It showed that the expansion of cultivated land was contributed by the reclamation of grassland.

IOP Conf. Series: Earth and Environmental Science **74** (2017) 012011 doi:10.1088/1755-1315/74/1/012011

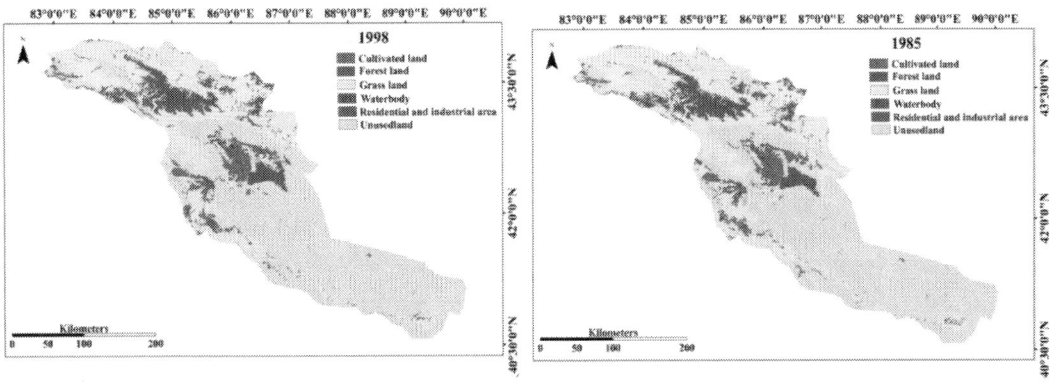


**Figure.2** Land uses and land cover interpretation map of BLB in 1985, 1998, 2008 and 2015.

*3.2 Determination of the Driving forces for the LUCC in the BLB*

It is apparent that both natural factors and human activities, and most likely their interactions were being the driving forces for tempo-spatial changes of the LUCC in Bosten Lake Basin from 1985 to 2015. Climatic conditions ~~have~~ play restraining role in utilization of land, which mainly shown in the distribution and composition of topography, grassland and forest, water resource, farming system and yield. Compared with the natural driving forces, human activities may have concentrated and immediate impacts on LUCC, whose effects usually take relatively long times to become noticeable.

In order to identify influential factors of LUCC change in BLB, we selected important factors to find out the main components of LUCC. For identifying the driving forces of LUCC in the BLB, this research take the cultivated land and residential land change quantity as the main changed land use types. According to PCA method, nine analysis factors are selected such as irrigation volume ($X_1$), runoff volume ($X_2$), temperature ($X_3$), precipitation ($X_4$), irrigated area ($X_5$), GDP ($X_6$), Per capita GDP ($X_7$), population ($X_8$) and primary industry ($X_9$) etc.

The results of principal component analysis revealed that the accumulative contribution rate of the first and secondary principal components is 87.9% (Table. 2), which is larger than 85% threshold value of accumulative contribution rates. Thus, it is fit with the principal component analysis requirement. Therefore, most variation could be explained by two principal components. The accumulative contribution percentage of the first principal component is larger than the secondary principal component, and the first principal component has controlled the most of the changes in the whole changing process. The results showed that there is an obvious positive correlation between the first principal component and observation variables $X_1$, $X_5$, $X_6$, $X_7$, $X_8$ and $X_9$, and the correlation coefficients are more than 0.9. Similarly, there is significant a positive correlation between the secondary principal component and the observation variables $X_2$, $X_3$, $X_4$, and correlation coefficients are over 0.83.

**Table 2.** Contribution rate of characteristic values and principal components.

| Component | Eigen values | % of Variance | Cumulative % |
|---|---|---|---|
| 1 | 6.308 | 63.084 | 63.084 |
| 2 | 2.482 | 24.820 | 87.904 |

**Table 3.** (Principal component) matrix.

| Component | irrigation volume | runoff volume | temper-ature | precipi-Tation | irrigated area | GDP | percapita GDP | popul-ation | primary industry |
|---|---|---|---|---|---|---|---|---|---|
| 1 | 0.954 | 0.119 | -0.269 | 0.260 | 0.911 | 0.958 | 0.972 | 0.981 | 0.969 |
| 2 | 0.060 | 0.905 | 0.830 | 0.833 | 0.077 | 0.321 | 0.114 | 0.018 | 0.068 |

From the Table 3, it can be indicated clearly that the first principal component which influences the change of cultivated land and residential land in BLB over the past thirty years shall be human activities. Those human activities include mainly the population growth and economic development. Therefore, it is determined that the first principal component is Socio-economic factors. The secondary principal component is composed by three factors and the sequence of their impact can be ordered as runoff volume > precipitation > temperature. The second principal component is related with hydro-climatic factor, which is defined as the natural factors.

## 4 Quantities analysis of driving force in BLB

### *4.1 Analyzing the influence of Socio-economic factors*

PCA results indicated that, first principal component contains 63% of the variation rate, it is the main principal controlling factor and it can control the change of the other principal component. In the space, the first principal component express the social progress and the regional development process, as well as the impacts of human activities on the development of natural resources.

#### *4.1.1 Impact of population growth on LUCC.*

Population growth can be an essential driving force for LUCC in the study area, and meanwhile, the dynamics of population can be a response to the change of environment (Zhao et al., 2013). The population of the BLB was $58.3 \times 10^4$ in 1985, $69 \times 10^4$ in 1990, $99 \times 10^4$ in 2000, $121.3 \times 10^4$ in 2010, and $146.8 \times 10^4$ in 2015 (Xinjiang Uygur Autonomous Region Years Book, 1985～2015). The population growth directly resulted in expansion of socio-economic related land uses . First, the growing population required more land to meet basic needs of living, including food and place of residence. Second, the population's desire for economic development demanded more land for producing commodities. Particularly in this area, a significant proportion of cultivated land has been devoted to the profitable cotton production. Figure. 3 shows the correspondence between the growths of population and cultivated land in the BLB during 1985～2015. The correlation coefficient between the population and cultivated land is 0.942. The cultivated land had a slow growth from 1985～2000, and then had a jump during 2000～2015. With the continuous growth of population, large areas of wasteland and natural grassland had been reclaimed for cultivation. The area of cultivated land increased during 1985～2015, at the same time the population maintained a rising trend.

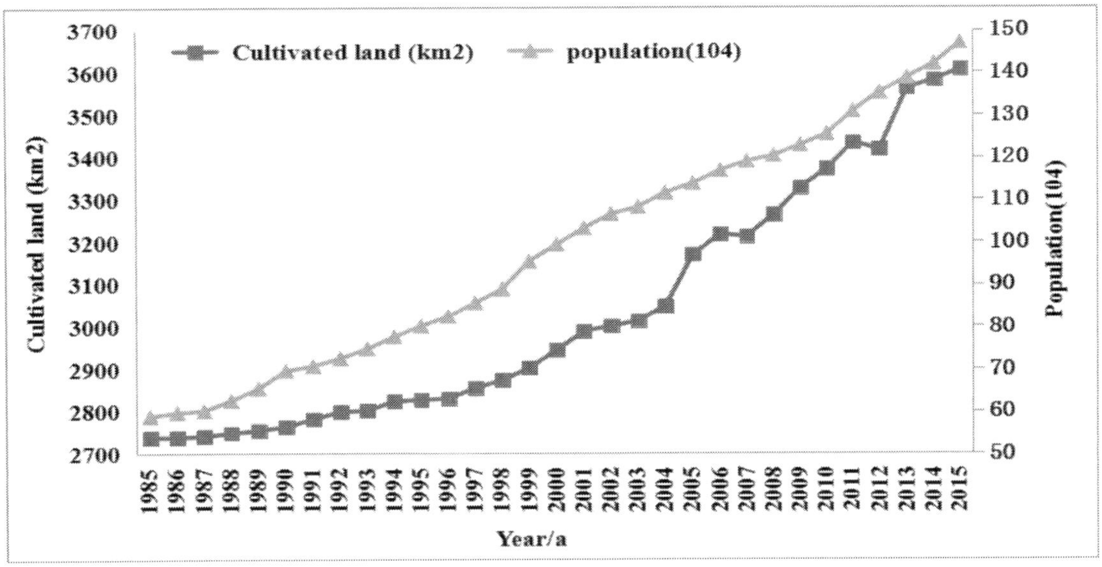


**Figure.3** Changes of population and cultivated land in the BLB during 1985~2015.

*4.2 Analyzing the Influence of natural factors*

In BLB, the LUCC is largely controlled by the volume and spatial distribution of water. In addition, climate factors including temperature and precipitation has a considerable effects on the LUCC. The mountain region of BLB is the characterized with abundant precipitation. The mountain areas are the formation zone of runoff. Then, the runoff flow into oasis and the desert region in the study area. Under global climate change, the runoff in the BLB has been experiencing a drastic change, and it is also a major driving factor for the LUCC.

4.2.1 *The climate changes.*

Figure. 4 shows the temporal variation of climate in the BLB during 1985~2015. The temperature of the study area had an increasing trend, especially in recent two decades (1995~2015). The average annual temperatures during 1995~2015 (8.8 ℃) are higher than those during 1985~1994 (8.71 ℃). In past three decades, the precipitation also had a slightly increasing trend. The average annual precipitation for 1995~2015 (140.5 mm) is more than that for 1985~1994 (114.21 mm). On the one hand, the increases of precipitation in the study area were certainly beneficial to both the human system and natural vegetation in those areas, alleviated the competition between the two systems for water. On the other hand, the increased precipitation and the warming trend in the mountain region that had accelerated the melting of snow and glaciers had resulted in increased runoff, which bring positive effect for expansion of cultivated land. However, the extra water from the upstream areas in the form of runoff has given humans more control in the use and allocation of the water.

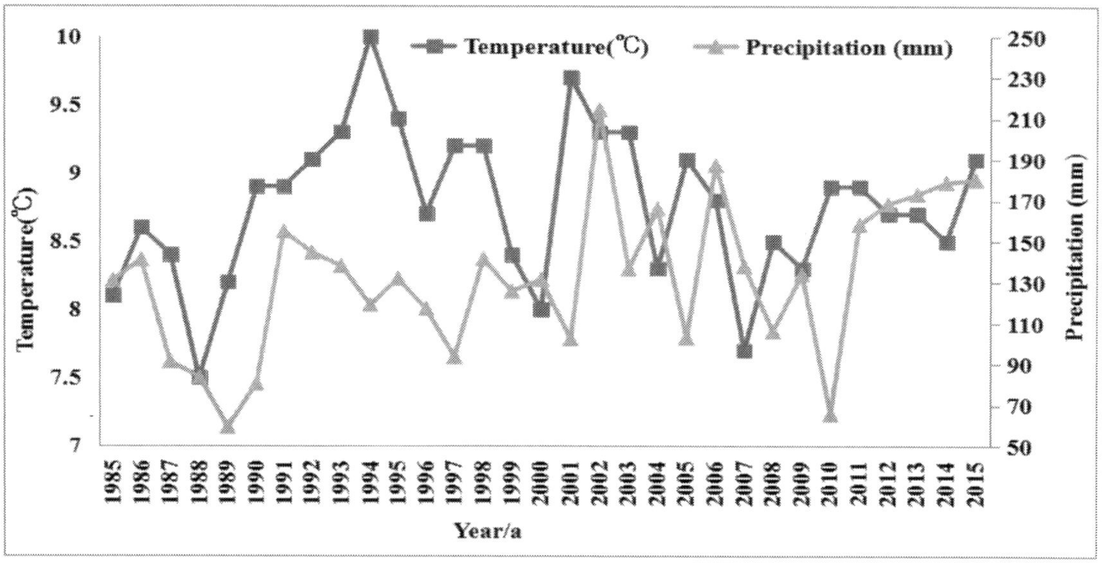


**Figure.4** Changes of Climate in the BLB during 1985～2015.

4.2.2 *Hydrological change.*

Figure. 5 shows that the runoff of the Kaidue River had an increasing trend during the past three decades. In the most recent two decades, the average annual runoff of the Kaidue River in 1985, 1995, 2005, and 2015 are $29.8 \times 10^8$ m$^3$, $34.18 \times 10^8$ m$^3$, $42.59 \times 10^8$ m$^3$, and $42.8 \times 10^8$ m$^3$, respectively, with the amount of the second 10 years is greatly larger than that of the first 10 years. The average annual runoff of 1995～2015 is $39.9 \times 10^8$ m$^3$, a 3.2 % increase compared with that of 1985～1994.It is apparent that, the runoff of the northwest arid area strongly depends on glaciers. Glacier change has a significant impact on water resources change in the northwest arid area. The influence of temperature on Glacier is mainly expressed in the change of melted water, and in this way the temperature change result in the change of runoff amounts the arid region.

## 5. Conclusions

From the results, the following conclusions can be drawn:(1) From the four inch of remote sensing image interpretation, it was found that From 1985 to 2015 as the expanding land use types cultivated land, forest land, residential land and unused land increased 31.7%, 9.7%, 1.2%, 86.68%, 1.6% respectively, while the areas of grassland decreased 7.2% in the same period; (2) The result of PCA indicated that the observed LUCC have been caused by a combination of human activities and climate factor, and the first principal component contains 63% of the variation rate, it is the principal controlling factor and it can control the change of the other principal component;(3) the second component contains 24.8 % of the variation rate, it is also one of the main component which significantly contribute to changes the LUCC.

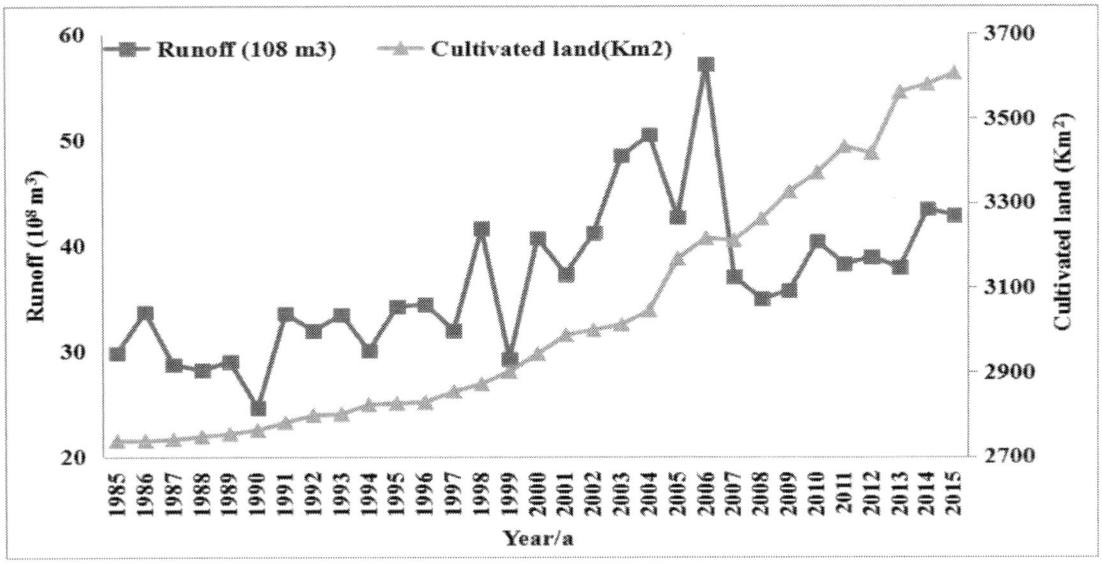


**Figure.5** Changes of Runoff and cultivated land in the BLB during 1985～2015.

**Acknowledgements**

This research was supported by Director Youth Fund Project of Institute of Remote Sensing and Digital Earth CAS, (Y6XS6000CX) and Western light Project of CAS (2015-XBQN-A-07).

**References**

[1] Adel S, Ryutaro T 2007 Remote sensing and GIS for mapping and monitoring land cover and land - use changes in the Northwestern coastal zone of Egypt *Appl. Geogr.* **27** 28–41

[2] Adenew T.A, Jaspervan V, Peter H V 2015 Land - use and land - cover changes in the Central Rift Valley of Ethiopia: Assessment of perception and adaptation of stakeholders *Appl. Geogr.* **65** 28–37

[3] Chen J H, Zhang E L, Stephen J B, Huang X Z 2014 Relationships between chironomids and water depth in Bosten Lake, Xinjiang, northwest China *J. Paleolimnol.* **51**, 313–323.

[4] Chen Y, Ye Z, Shen Y 2011 Desiccation of the Tarim River, Xinjiang, China, and mitigation strategy *Quatern. Int.* **244** 264-271

[5] Deborah R., Rafael M S, Juan M T 2016. Estimation of human induced disturbance of the environment associated with 2002, 2008 and 2013 land use / cover patterns in Mexico. *Appl. Geogr.* **66** 22-34

[6] Zhao R F, Chen Y N 2013. Land use and land cover change and driving mechanism in the arid inland river basin: a case study of Tarim River, Xinjiang, China. *Environ. Earth Sci.* **68** 591-604.

IOP Conf. Series: Earth and Environmental Science **74** (2017) 012012 doi:10.1088/1755-1315/74/1/012012

# The spatial distribution of remote sensed $SO_2$ in China and its relationship with energy consumption

**Chao Xu[1, 2, 3], Yunpeng Wang[1*], Lili Li[1], Yihong Liu[4]**
1 Guangzhou Institute of Geochemistry, Chinese Academy of Sciences, Guangzhou 510640, China
2 University of Chinese Academy of Sciences, Beijing 100049, China
3 Guangzhou Institute of Energy Conversion, Chinese Academy of Sciences, Guangzhou 510640, China
4 School of Geographic and Oceanographic Sciences, Nanjing University, Xianlin Road No.163, Qixia District, Nanjing 210046, China

E-mail: wangyp@gig.ac.cn

**Abstract.** The spatial distribution of $SO_2$ in China was analyzed using OMI-Level3 $SO_2$ PBL (planetary boundary layer) column concentration data of 2013. At the same time, the total value of energy consumption per unit area and coal energy consumption per unit area were analyzed and compared with $SO_2$ distribution. Results show that $SO_2$ distribute unevenly in whole China, and even at the provincial level, $SO_2$ concentration difference was obvious. The spatial distribution of $SO_2$ in China displays characteristics of two highest plates as the center, sub-high regions radiate from center, as well as "The east higher than the west, and the north higher than the south". Energy consumption per unit area and coal consumption per unit area also distribute unevenly, and the two distributions have similar characteristics but different details. The higher similar spatial distributions of $SO_2$ and energy consumption (especially coal consumption) imply the firm connection of $SO_2$ pollution and coal consumption. $SO_2$ pollution in China remains acute, and the control on coal consumption remains important and emergent in recent years.

## 1. Introduction

Sulphur dioxide ($SO_2$) is one important gaseous pollutant in the troposphere [1], which not only impacts air quality, but also human health and environment through the forms of acid rain or fine particle [2-3]. The source of atmospheric $SO_2$ includes anthropogenic and natural ones. The burning of fossil fuels in human activity is the main source of $SO_2$ [4-5]. Especially in China, as economic development and fossil fuel consumption increasing, monitoring and research of $SO_2$ are necessary. Most traditional research used data from $SO_2$ monitoring stations, which confine the scale of study time and space [6-7]. China is a vast country, and regional difference exists in many affects, including economy, energy consumption and $SO_2$ distribution. Therefore, this article analyses the spatial distribution of $SO_2$ in China and its relationship with energy consumption using the remote sensed and statistical data.

## 2. Data Processing and Method

* Corresponding author.

*2.1 Data source and processing*

At present, remote sensed $SO_2$ data are mostly from GOME, SCIAMACHY, OMI, GOME-2, and OMPS. Many researches [8-9] suggest retrieving concentration of trace gases by using OMI data achieved higher precision. The data used in this study mainly include OMI-Level3 $SO_2$ PBL(planetary boundary layer) column concentration data with 0.25°× 0.25 °spatial resolution of 2013, which were downloaded from NASA website (http://acdisc.gesdisc.eosdis.nasa.gov/). $SO_2$ column concentration of planetary boundary layer is closely related to human activities. And OMI-Level 3 $SO_2$ PBL data are filtered from OMI-Level 2 $SO_2$ PBL column concentration data according to observation angel, abnormal row and other conditions.

Energy consumption data of provinces in China (given the difficulty in data acquisition, Hong Kong, Macao, Taiwan, and Tibet were not included) of 2013 used in the study area were from China energy statistical yearbook. China and provinces boundary vector data are downloaded from National Geometrics Centre of China.

*2.2 Method*

In order to quantitatively analyze the distributions of $SO_2$ concentration, energy consumption and coal consumption, this article introduced spatial correlation matrix of Geoinformatics to measure the correlation. Spatial correlation matrix is used to calculated the correlation coefficient of data which have the same space range [10], formula is shown in formula 1,the value range is [-1,1]. While 1 represent completely correlation, -1 represent completely negative correlation.

$$Corr_{ij} = \sum_{k=1}^{n} \frac{(v_{ik} - \mu_i)(v_{jk} - \mu_j)}{(n-1)\sigma_i \sigma_j} \tag{1}$$

$Corr_{ij}$——the correlation coefficient of the distributions of $SO_2$ concentration and energy consumption (coal consumption), $i$—the distributions of $SO_2$, $j$—the distributions of energy consumption (coal consumption), $\mu$——mean value of raster data, n——number of pixels, $v_{ik}$——the $k$ th pixel value of $i$ layer, $v_{jk}$—— the $k$ th pixel value of $j$ layer; $\sigma$ ——standard deviations of data layer.

## 3. Results and discussions

The spatial distribution of remote sensed $SO_2$ column concentration of China in 2013 was shown in Figure 1, where the $SO_2$ concentration of China can be divided into five levels. The highest $SO_2$ concentration area is located in two plates. One is the middle and lower reaches of the Yellow River Basin area, including the east of Shanxi, the south of Hebei, Tianjin, the north of Henan, Shandong and the north of Jiangsu. The other part is the Sichuan Basin area, including the southeast of Sichuan, the west of Chongqing and Guizhou. $SO_2$ concentrations in these areas were mostly higher than 0.6 DU. The radiation zones of the two plates in the first level and Pearl River Delta region constituted the second level of $SO_2$ concentration. Heilongjiang, the west of Jilin, the north of Liaoning, the east of Inner Mongolia, the southeast of Gansu, the east of Hunan, the south of Hubei, most part of Jiangxi, and the north of Guangdong concentrated in the third level. The forth level included Yunnan, Fujian, the north of Xinjiang and the west of Sichuan. Besides all above areas, the rest areas fall in the fifth level, which mostly located in the northwest of China.

Figure 2 and 3 show the spatial distributions of energy consumption per unit area and coal consumption per unit area of China in 2013, respectively. As shown in the Figures, just like $SO_2$ distribution, energy consumption per unit area and coal consumption per unit area distributed unevenly. The two distributions have similar characteristics but different details. The high energy consumption per unit area zones focus on Bohai Bay Economic Region (Beijing, Hebei, Tianjin, Liaoning), east coast region (Jiangsu, Zhejiang, Shanghai) and Guangdong. The sub-high region including Shanxi, Henan, Ningxia, followed by Shannxi, Hubei, Anhui, Guizhou, Hunan, Jilin and Fujian. The rest provinces have relatively low energy consumption per unit area.

Comparing with distribution of energy consumption, coal consumption's high regions focused on the Yellow River Basin area and east coast regions, including Shanxi, Hebei, Tianjin, Henan, Shandong, Jiangsu and Shanghai. Beijing, Liaoning, Anhui, Zhejiang, Ningxia and Guangdong composed the sub-high region. The third high coal consumption per unit area region located in Shannxi, Chongqing and Guizhou, followed by Hubei, Hunan and Jiangxi. The rest provinces constituted the low value zone.

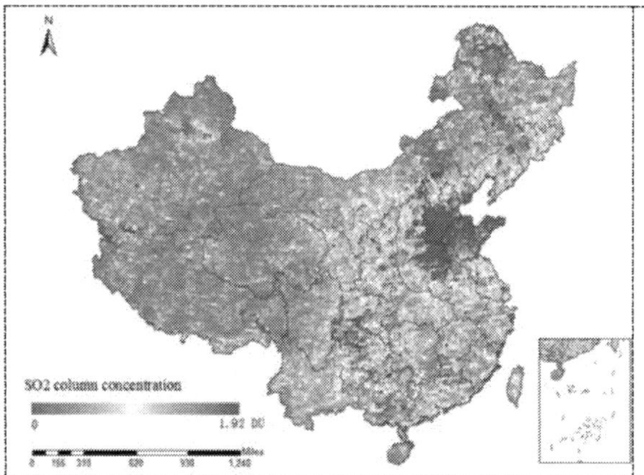


**Figure 1.** The spatial distribution of remote sensed $SO_2$ column concentration of China in 2013.

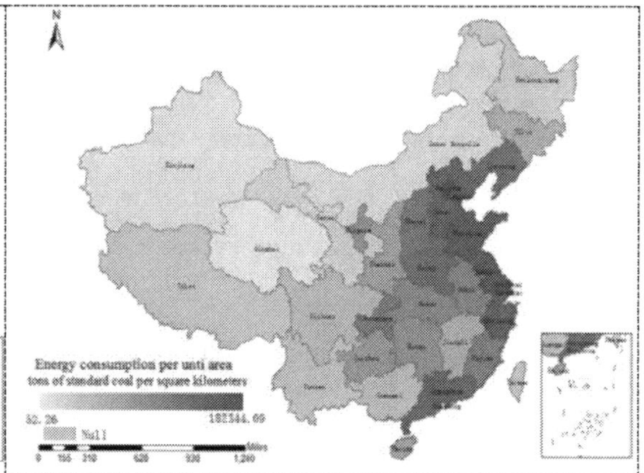


**Figure 2.** The spatial distribution of energy consumption per unit area of China in 2013.

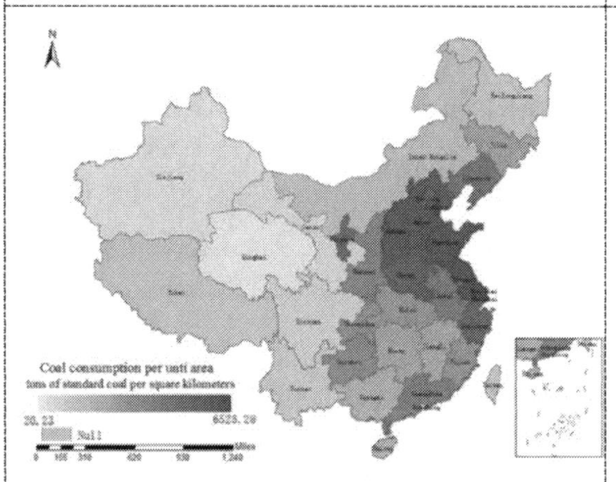


**Figure 3.** The spatial distribution of coal consumption per unit area of China in 2013.

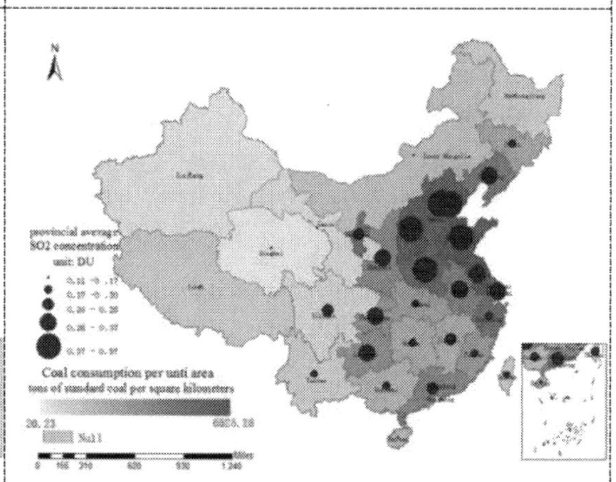


**Figure 4.** The overlapped distributions of provincial average $SO_2$ concentration and coal consumption per unit area of China in 2013.

Table 1 gives the correlation coefficient calculated result of the distribution of $SO_2$ and the spatial distribution of energy consumption and coal consumption. The correlation coefficient of $SO_2$ and energy consumption is 0.6578, while the coefficient with coal consumption reached 0.8707. It is easy to conclude that the spatial distribution of $SO_2$ does have connection with the spatial distribution of energy consumption and coal consumption, showing higher similarity with coal consumption per unit area.

**Table 1.** The correlation coefficient of distribution of $SO_2$ and energy consumption

| | energy consumption per unit area | coal consumption per unit area |
|---|---|---|
| $SO_2$ | 0.6579 | 0.8707 |

It's worth noting that some provinces' coal consumption distribution and average $SO_2$ distribution appear inconsistent, especially reflected in Zhejiang and Sichuan. Zhejiang belongs to the high region of coal consumption per unit area while belong to the fourth level of provincial average $SO_2$ concentration. The southeast of Sichuan located in one of the highest $SO_2$ concentration plate while its provincial average $SO_2$ concentration declined markedly, only was 0.21DU. The inconformity mainly and directly caused by the non-uniformity of $SO_2$ concentration distribution has been ignored in provincial scale. If the influence could be eliminated, consistency of $SO_2$ concentration distribution and coal consumption per unit area distribution would be higher, though at present they have shown high similarities.

## 4. Conclusion

In this paper, the spatial distribution of $SO_2$ in China was analyzed using OMI-Level3 $SO_2$ PBL (planetary boundary layer) column concentration data of 2013. $SO_2$ distributes unevenly in whole China, even at the provincial level, and $SO_2$ concentration difference was obvious. $SO_2$ concentration of China can be divided into five levels, the highest $SO_2$ concentration area located in the middle and lower reaches of the Yellow River Basin area and Sichuan Basin area, where $SO_2$ concentration were mostly higher than 0.6 DU. The radiation zone of the first level region and Pearl River Delta region constituted the second level of $SO_2$ concentration. In summary, the spatial distribution of $SO_2$ in China displays characteristic of "two highest plates as the center, sub-high regions radiate from center; and the east being higher than the west, the north being higher than the south". Energy consumption per unit area and coal consumption per unit area also distributed unevenly, and the two distributions have similar characteristics but different details. The spatial distribution of $SO_2$ do has connection with the spatial distribution of energy consumption and coal consumption, and show higher similarity with coal consumption per unit area. $SO_2$ pollution in China remains acute, and the controi of coal consumption remains important and emergent in recent years.

## Acknowledgements

China 863 Program (2006AA06A306) is acknowledged for financial supports.

## References

[1] Wang M X 1999 *Atmospheric Chemistry* (Beijing: China Meteorological Press) pp 30-35

[2] Wang Q X, Liu Y and Pan X C 2008 Atmosphere pollutants and mortality rate of respiratory diseases in Beijing *Sci. Total Environ.* **39** 143-8

[3] Dai S G 1997 *Environmental Chemistry* (Beijing: Higher Education Press) pp 88-9

[4] Reddy M.S and Venkataraman C 2002 Inventory of aerosol and sulphur dioxide emissions from India: Fossil fuel combustion *Atmos. Environ.* **36** 677-97

[5] Giannitrapani M, Bowman A,Scott E,and Smith R 2007 Temporal analysis of spatial covariance of $SO_2$ in europe from 1990 to 2001 *Environmetrics* **18** 409-20

[6] Wang K 2008 *Analysis on the spatio-temporal gradients of carbon monoxide in China using remotely sensed data during 2003-2005* (Beijing: Nanjing University)

[7] Cheng L G, Wang Y J and Wang Y T 2005 Summarization of the application of remote sensing in atmospheric environmental monitoring *Environmental monitoring in China* **21** 17–23

[8] Echardt S, Prata A J, Seibert P and Stebel K 2008 Estimation of the vertical profile of sulphur dioxide injection into the atmosphere by a volcanic eruption using satellite column measurements and inverse transport modelling *Atmosphere Chemistry and Physics Discussions* **8** 3761-805

[9] Xiao Z Y and Jiang H 2011 Comparison of total ozone derived from OMI and SCIAMACHY over Asia region *China Environmental Science* **31** 529-39 (In Chinese with English Abstract)

[10] Shi Y Q, Wang Q and Zhang Z P 2014 Analysis on spatial distribution of air pollution and its spatial correlation with influencing factors in Xiamen City *Chinese Journal of Environmental Engineering* **12** 5406-12

IOP Conf. Series: Earth and Environmental Science **74** (2017) 012013 doi:10.1088/1755-1315/74/1/012013

# Analysis on dry-wet characteristics in the Zhujiang River Basin in recent 55 years

**Jizhong Qiu**[1,2], **Yunpeng Wang**[1*] **and Jie Xiao**[1,2]

1 Guangzhou Institute of Geochemistry, Chinese Academy of Sciences, Guangzhou 510640, China
2 University of Chinese Academy of Sciences, Beijing 100049, China

E-mail: wangyp@gig.ac.cn

**Abstract**. Based on Palmer Drought Severity Index (PDSI), this paper analysed the spatial distribution of dry-wet condition in the Zhujiang River Basin in 1961-2015, and studied the dry-wet climatic regionalization by Rotated Empirical Orthogonal Function (REOF). The results show most areas in the Zhujiang River Basin was in normal state of dry-wet condition and the whole basin can be divided into 7 dry-wet climate zones. The above conclusions reflect the dry-wet characteristics in the basin reasonably, which can provide a reference on prediction, monitoring and evaluation of some kinds of extreme climatic disasters such as floods and droughts inextricably linked with dry-wet condition.

## 1. Introduction

In 2013, the Fifth Assessment Report that was published by the Intergovernmental Panel on Climate Change (IPCC) thought that warming of the climate system is unequivocal [1]. Under the global warming background, various extreme climatic events occurred in the Zhujiang River Basin [2-3]. Hence, production and living in the region has been seriously affected.

At present, there are a lot of researches on dry-wet climatic regionalization in the whole Zhujiang River Basin. Most researchers thought the basin is a humid climatic region [4]. However, there are little understanding about the distribution and division of dry-wet condition in the interior basin, which is regional essentially important. Therefore, it is necessary to analyze the characteristics of dry-wet events in the basin and study the occurrence regularity from the perspective of regional differences.

The annual PDSI values in 1961-2015 were selected to analyze the spatial distribution of dry-wet condition in the Zhujiang River Basin. Simultaneously, based the spatial patterns of REOF, the interior basin was divided. These results can provide theoretical basis about the understanding of dry-wet characteristics in the area and forecasting and relieving extreme climatic events.

## 2. Study area, data and methods

### *2.1. Study area and datasets*

The Zhujiang River is an important river in the Southern China. It is composed by the Xijiang River, the Dongjiang River, the Beijiang River and the Delta Rivers. The total length of the river is about

[*] Corresponding author.

2400 km, and the total area of its basin is about 4.6×105 km2. The Xijiang River Basin is further divided into four sub zones: Nanbeipan, Guibei, Guizhong and Yuegui in this paper. The basin belongs to subtropical monsoon region. In the whole basin, the mean annual temperature is between 14-22C° and the mean annual precipitation is about 1525.1mm that is markedly reduced from east and west. Moreover, the distribution is uneven in one year time and there are big differences in regional distribution and inter annual variation. [5]

The weather data that was used to calculate PDSI is from China Meteorological Data Website (http://data.cma.cn). There are 45 stations in all and the time period is from 1961 to 2015. The soil data is from Harmonized World Soil Database of the Food and Agriculture Organization (http://www.fao.org).

### *2.2. Methods*

#### *2.2.1. Calculation of Palmer Drought Severity Index (PDSI)*

PDSI can well represent regional dry-wet condition and is suitable for dry-wet monitoring and evaluation in different time scales [6]. Considered the impact of multiple factors of temperature, precipitation and soil type, the index was established by water supply and demand. In 2006, PDSI became the national standard in China after correction. The calculate steps can be found in [7]. The dry-wet classification is shown in Table 1. The classifications (Table 1) shows PDSI<=-1.0 indicates dry while PDSI>=1.0 indicates wet.

**Table 1.** Dry-wet classification according to PDSI

| Category | Index value |
|---|---|
| extreme wet | PDSI>=4.0 |
| severe wet | 4.0>PDSI>=3.0 |
| moderate wet | 3.0>PDSI>=2.0 |
| mild wet | 2.0>PDSI>=1.0 |
| normal | 1.0>PDSI>-1.0 |
| mild dry | -2.0<PDSI<=-1.0 |
| moderate dry | -3.0<PDSI<=-2.0 |
| severe dry | -4.0<PDSI<=-3.0 |
| extreme dry | PDSI<=-4.0 |

#### *2.2.2. Rotated Empirical Orthogonal Function (REOF)*

Empirical orthogonal function (EOF) decomposes an original variable field X into spatial eigenvectors V, and associated time coefficients T (Eq.1) [8]. Meanwhile, EOF extracts the main information from X on some few eigenvectors in favor of analysis and processing.

$$X_{m\times n} = V_{m\times p} T_{p\times n}, \tag{1}$$

In this study, the REOFs are derived from the PDSI, which convert the dry-wet information into the rotated eigenvectors with both spatial and temporal significances. On the basis of EOF, REOF rotates eigenvector V further which makes the high loadings derived from the spatial patterns (denoted by EOFs in this paper) only exist in a certain region. It highlights the areas with high values in a smaller range which is easier to identify the spatial pattern. There is a high correlation between each spatial range and a loading vector after rotation. The dry-wet characteristics after rotation is more stable than that before rotation. So, the loading vectors can better represent spatial anomaly characteristics of dry-wet change. VARIMAX criterion of orthogonal rotation was selected to handle variable field PDSI. Detailed steps and formulas can be seen in [9].

## 3. Results and analysis

The spatial distribution of dry-wet condition in the Zhujiang River Basin (Figure 1) shown, most areas were in normal state of the dry-wet condition in 1961-2015. There are no obvious High value centers, especially in Nanbeipan Zone and Dongjiang Basin. The four low value centers of dry-wet distribution are Western Guibei Zone, Central Guizhong Zone, Northern Yuegui Zone, and the border area among Beijiang Basin, Xijiang Basin and Delta. The low centers are all in mild dry state. The dry-wet contours that locals in the neighbor of the stations in above areas are more concentrated, illustrates dry-wet spatial changes fluctuate dramatically.

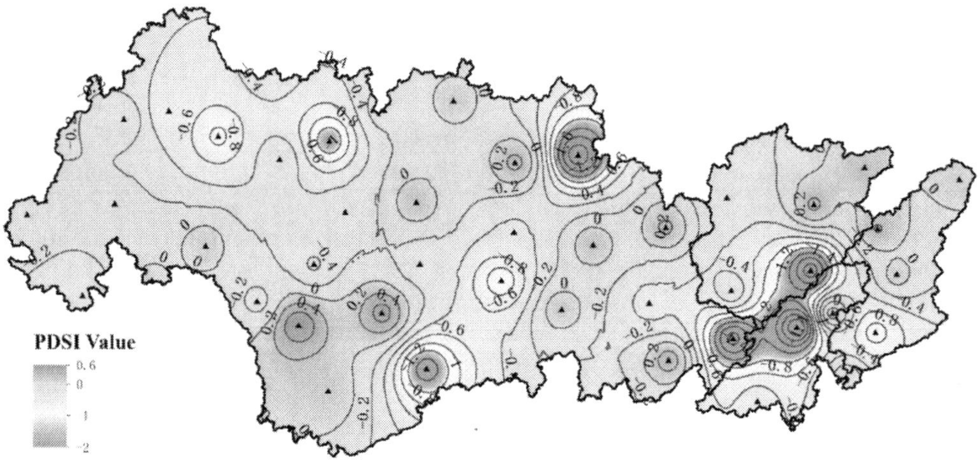


**Figure 1.** The spatial distribution of dry-wet condition in the Zhujiang River Basin in 1961-2015

The distribution of dry-wet condition indicates there are some big differences in various areas. Therefore, this study divided the Zhujiang River Basin by REOF. In REOF1 derived from dry-wet variable field, the positive extreme areas of rotated loading vector are locating in Southern Nanbeipan Zone, Eastern Guizhong Zone and Central Yuegui Zone. The negative extreme areas are locating in Western Yuegui Zone and North-eastern Beijiang Basin. It represents the distribution pattern where the dry-wet trend in the Central Basin was opposite to the Eastern and Western Basins. In other words, the Central Basin was wet while the Eastern and Western Basins were dry, and vice versa. In REOF2, the positive areas are locating in Southern Nanbeipan Zone and Northwestern Yuegui Zone. The negative areas are locating in Northern Guibei Zone, Southern Yuegui Zone and Central Dongjiang Basin. It shows the distribution pattern in which the dry-wet trend in the Western Basin was opposite to the Eastern and Central Basins. In REOF3, the positive areas are locating in Delta and Southern Dongjiang Basin. The negative areas are locating in Central and Eastern Nanbeipan Zone and Western Guizhong Zone. It represents the distribution pattern in which the dry-wet trend in the Eastern Basin was opposite to the Central and Western Basins. In REOF4, the positive areas are locating in North-eastern Yuegui Zone, North-western and South-eastern Beijiang Basin and Central Dongjiang Basin. The negative areas are locating in Southern and Eastern Nanbeipan Zone. It means the distribution pattern in which the dry-wet trend in the North-western Basin was opposite to the other areas in the Basin. In REOF5, the positive areas are locating in Northern Nanbeipan Zone, Central Yuegui Zone, South-eastern Beijiang Basin and North-eastern Dongjiang Basin. The negative areas are locating in South-western Nanbeipan Zone, Central Guizhong Zone, South-eastern Guibei Zone and North-western Yuegui Zone. It implies the distribution pattern the dry-wet trend in the Northern Basin was opposite to the Eastern and Southern Basins. In summary, according to the distribution of the high loading areas in the first 5 REOFs, the Zhujiang River Basin is divided into 7 dry-wet climatic zones A-G (Figure 2).

## 4. Conclusion

Based on above results, this study found that:

1. In recent 55 years, Western Guibei Zone, Central Guizhong Zone, Northern Yuegui Zone, and the border area among Beijiang Basin, Xijiang Basin and Delta are in mild dry states. Meanwhile, the other areas are in normal state of dry-wet condition.

2. The Zhujiang River Basin can be divided into 7 various dry-wet climatic regions. To the other regional differences of dry-wet condition, for example, tendency, mutation and period, we need have further researches.

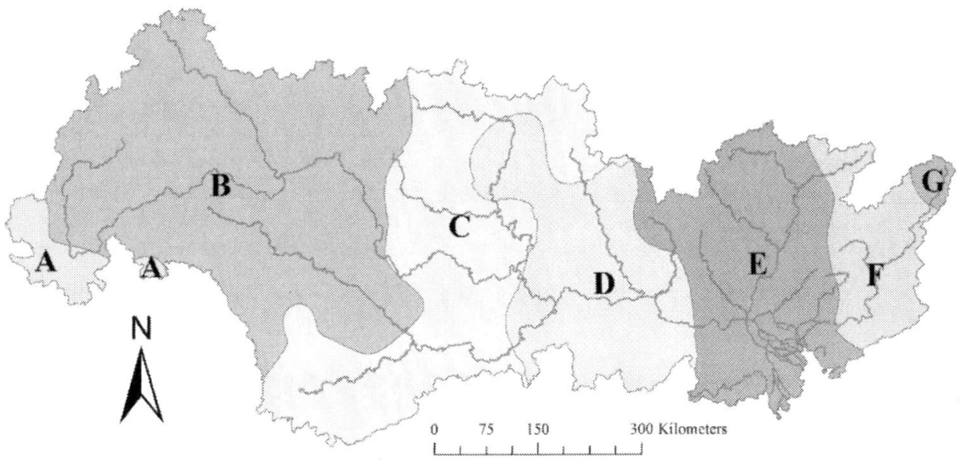


**Figure 2.** Climatic regionalization of dry and wet change in the Zhujiang River Basin

**Acknowledgements**

China NSFC (41401485) and 863 Program (2006AA06A306) are acknowledged for financial supports.

**References**

[1] IPCC Climate Change 2013 *The Physical Science Basis. Contribution of Working Group I to the Fifth Assessment Report of the Intergovernmental Panel on Climate Change* (Cambridge, United Kingdom and New York, NY, USA: Cambridge University Press) p 1535

[2] Fischer T, Gemmer M, Liu L L 2012 Change-points in climate extremes in the Zhujiang River Basin, South China, 1961-2007 *Climatic Change* **110**: 783-99

[3] Wu C H, Huang G R 2016 Projection of climate extremes in the Zhujiang River basin using a regional climate model *Int. J. Climatol.* **36** 1184-96

[4] Wu S H, Yin Y H, Zheng D 2005 Study on dry-wet surface condition of China in recent 30 years *Sci. China, Ser. D: Earth Sci.* **35** 276-83

[5] Li X L, Yang T 2015 Spatio-temporal Changes Prediction of Precipitation and Temperature in Pearl River Basin from 2015 to 2100 *Water Resources and Power* **11** 5-9 (In Chinese with English Abstract)

[6] Palmer W 1965 *Meteorological drought* (U.S. Department of Commerce Weather Bureau)

[7] Chinese National Climate Center 2006 *Classification of meteorological drought* (Beijing: Standardization Administration of the People's Republic of China)

[8] Edward N 1956 Lorenz Empirical Orthogonal Functions and Statistical Weather Prediction *Sci. Rep.* **409**: 997-99

[9] Huang J Y2004 *Climatic Statistical Analysis and Forcasting Methods* (Beiing: China Meteorological Press)

IOP Conf. Series: Earth and Environmental Science **74** (2017) 012014 doi:10.1088/1755-1315/74/1/012014

# Application of remote sensing information related to mineralization alteration and new discovery in geological prospecting work

**F J Yao[1], B H Fu[2]**

1 MLR Key Laboratory of Metallogeny and Mineral Assessment, Institute of Mineral Resources, Chinese Academy of Geological Sciences, Beijing, 100037, China
2.Institute of Remote Sensing and Digital Earth, Chinese Academy of Sciences, Beijing, 100094, China

E-mail: fjyao@sina.com

**Abstract.** As an intuitive manner, Remote Sensing (RS) can be used in resource prospecting and it shows some effects in practice, but high resolution and accuracy still remain problematic because RS methods need to be improved. Based on a RS prospecting model, prospecting resolution and accuracy can be greatly elevated. Different resources need to build various RS prospecting models. Some targets were proposed using RS exploration model. According to field examination, new discoveries have been successfully obtained in hydrothermal metal mineral prospecting, sedimentary metal mineral prospecting. Therefore the RS becomes more effective method in prospecting.

## 1. Introduction

RS(Remote Sensing) has received a very significant effect in geological prospecting [1-3]. Initially, some limonite alteration rocks were extracted using Landsat multispectral data in 1977 [4]. The spectral characteristics were collected about hydroxyl minerals in Visible Near-Infrared (VNIR) to Short Wave Infrared (SWIR) [5]. Remote sensing alteration anomalies were extracted using RS, specifically, Principal Components Analysis (PCA) was used to estimate the response of alteration minerals and was applied in many zones [6-7]. In 2007, China started-up a big project called 'Mineral Resources Potential Assessment Project', in which alteration RS anomalies extraction methods using Landsat TM/ETM were extended throughout the whole Chinese mainland regarded RS techniques as an important methods [8]. Landsat TM/ETM data could not be used to distinguish minerals or mineral assemblages according to less wave bands, but the Advanced Spaceborne Thermal Emission and Reflection Radiometer (ASTER) remote sensor provided a high spectral resolution remote sensing data, containing 14-bands: nine bands are in VNIR-SWIR regions and the other five bands are in Thermal Infrared (TIR) region. ASTER can be used for extracting alteration minerals and studying alteration zonation [9-10]. And thermal infrared RS (e.g. ASTER TIR) could be used to study the geothermal resources [11-12]. However, the Alteration RS anomalies just indicate alterations related to hydrothermal metal minerals, but can not be used in sedimentary ore deposits prospecting. In our research, based on extraction useful RS information, the RS prospecting models were built for studied hydrothermal metal minerals, sedimentary metal ores. Finally, some targets were listed and validated in many Chinese fields, with new discoveries.

IOP Conf. Series: Earth and Environmental Science **74** (2017) 012014 doi:10.1088/1755-1315/74/1/012014

## 2. Data and methodology

### *2.1. Study Area and Data*

In our work, some areas in China were chosen for prospecting the hydrothermal metallic minerals, sedimentary metallic deposits. Figure.1 showed the positions of different types of resources for the RS models and prospecting. We used different types of RS data to research specific problems included ASTER data, PALSAR data, and LANDSAT data. These data were bought from PASCO CHINA. Types of data functions and purposes are shown in table1. In practice work, each data must be used combined with other data, but one kind of data was main used and other data served as supplements.

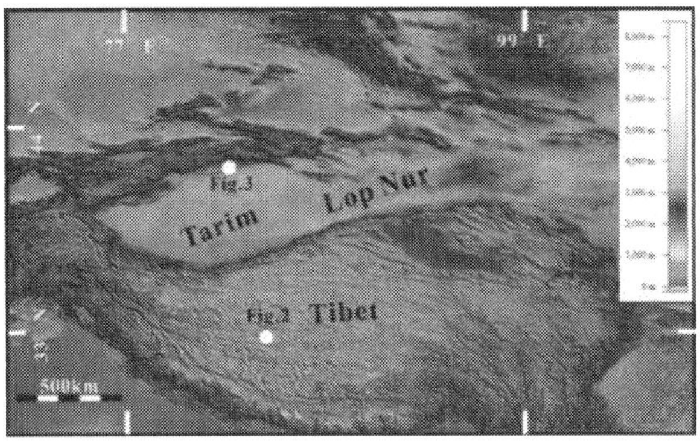


**Figure 1**. Locations of different types of resources for our research. The hydrothermal metallic minerals is located in Figure 2, the location of the sedimentary metallic deposits is marked on Figure3.

**Table 1.** Functions and purposes of RS data

| RS data | Resolution | Functions | Purposes |
|---|---|---|---|
| ASTER | 15m,30m 90m | Group-contained anomalies | Alteration mapping |
| PALSAR | 7m-44m | Structure information | Structural mapping |
| LANDSAT | 15m,30m 90m | Lithology information | Lithological mapping |

### *2.2. Methodology*

Image processing methods were chosen based on specific resources. For hydrothermal metal minerals, alteration zone RS information were main extracted. Electronic anomalies extraction and lithological recognition were used in sedimentary metal deposits. Tectonic information was extracted using radar data in overlay and coverage areas.

Firstly, ASTER data were used to study hydrothermal minerals. The extraction of the alteration zones and group anomalies through NIR-SWIR and TIR of ASTER data. We mainly used ratio and PCA methods of different bands. Secondly, weak lithology information of the sedimentary minerals could be revealed used ETM or ASTER images and electronic anomalies were extracted to indicate sedimentary minerals. Finally, Structural information interpreted by RADAR data (PALSAR). the RS model of resource explorated were built by overlying different information.

## 3. Image processing

IOP Conf. Series: Earth and Environmental Science **74** (2017) 012014 doi:10.1088/1755-1315/74/1/012014

### *3.1. The Hydrothermal Metallic Minerals*

In order to achieve the surface alteration zonation, about 30 samples were collected. Each sample was measure three times using ASD spectrometer and then each spectrum was analyzed for the alteration minerals distribution using PimaView software.

In Duolong deposits surface, we mainly collected the samples: metamorphic feldspathic quartz sandstone, granite porphyry, lithic quartz sandstone, gabbro, breccia, granite, basaltic andesite, quartz porphyry, and so on. Among them, 16 samples of Duobuza deposit and 18 samples of Bolong deposit were selected with sampling unequal interval along lithology and alteration variations. The alteration of the samples are limonitization, propylitization, chloritization, sericitization etc.

Major alteration zones of Duolong deposits include potassium silicification zone, phyllic zone, prophylitic zone, illite - hydromica zone and gossan. Alteration minerals have difference: K-feldspar, muscovite, chlorite, quartz in potassium silicification zone, mucovite, halloysite, illite, quartz in phyllic zone, calcite, chlorite in propylitic zone, illite, montmorillonite in illite-hydromica zone, and limonite in gossan. The spectral characteristics of all alteration minerals included SWIR(short-wave infrared) and TIR(thermal infrared) were analyzed and resampled to ASTER spectral responses(Figure 2).

**Figure 2.** RS alteration zones of Duolong ores, Tibet. According to spectral analysis, RS alteration zones information were extracted: potassic zone was shown in red, quartz-sericitization zone in yellow, propylitization zone in brown, illite-hydromica zone in pink, and gossan in blue. These anomalies of alteration zones show ring distribution.

The hydrothermal metallic minerals mostly have alteration zones, such as porphyry copper deposits. Firstly, alteration minerals assemblages of each alteration zone were analyzed, as well as spectral absorption characteristics of these minerals. Finally, alteration zones RS mapping were done using RS. Figure 2 shows the results of RS alteration zones in Duolong ore concentration regions, Tibet.

### *3.2. The Hydrothermal Metallic Minerals*

Image lithologic enhancement process was used to extract sedimentary metallic deposits information, such as sandstone-type copper deposits. lithologic information extraction are main methods.

The spectral characteristics of sandstone copper deposits and wall rocks are slightly different. Figure 3 show known Dishui Cu deposit position and ore-bearing bed. The spectra were collected from the position which number are 1, 2, and 3. The red spectrum was gathered from ore-bearing bed, which was difficultly distinguished, while black and green spectra gathered from wall rocks. We made the A-B profile. The spectra of ore-bearing bed were small subtleties contrasted ones of wall rocks.

Therefore, the special method was needed for enhancing lithologic information. In order to extract different lithologic characteristics, the frequency domain of the RS data need to re-analyze. In the result, the weak information were enhanced. Brown color indicated ore-bearing bed. Meanwhile, we extracted electronic anomalies. The lower anomalies were shown in sandstone copper layer.

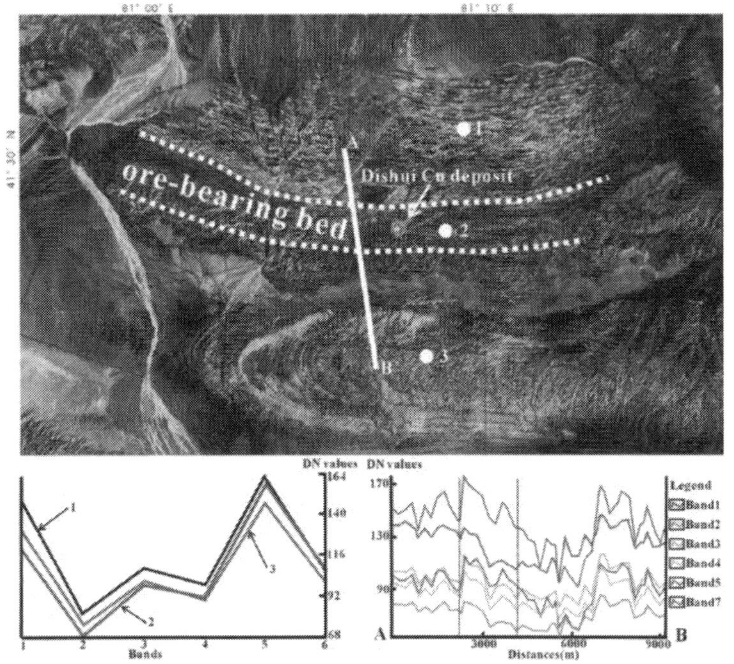


**Figure 3.**The known Dishui sandstone-type copper RS information in Baicheng, Xinjiang. The red spectrum collected from ore-bearing bed was difficultly distinguished with black and green spectra gathered from wall rocks. The spectra of ore-bearing bed were small subtleties compared ones of wall rocks form profile A-B.

## 4. Result

In order to build RS prospecting model, RS information must be extracted. More information contained geophysics, geochemistry, geology, and construction need to overlay RS information. And targets can be optimized.

### *4.1. Establishment of the RS Prospecting Model*

The RS prospecting model is based on RS information. Firstly, some important RS factors were chosen related to the analysis of the metallogenetic regularity. Secondly, RS information overlay geophysics, geochemistry, geology, and construction information on GIS platform. Finally, mineral prospecting indicators were established by analysis of different useful information and specific factors of resources exploration.

### *4.2. Application of RS Prospecting Model*

Through the RS prospecting model of sandstone copper deposits, many sandstone-type copper mineral deposits were discovered in Kuqa basin, Xinjiang. We found three copper mineralization layers. And the widths of the layers are 1.5m, 1.3m, and 10m. The 10m-wide copper mineralization layer was found in second field profile measurements.

### *4.3. Assessment*

Practical mineral exploration is effective using mineral RS prospecting model, but it is unable to find all mineral ores and mineralization occurrences using prediction of targets proposed by RS model. Therefore, it is inefficient using RS method in public. The RS was just regarded as "ornaments".

However, this job can improve the probability of RS prospecting. 1) RS is indeed a high-technical and practical method, which can be used to discovered new mineralization occurrences. 2) RS has many advantages, such as economy, environment protection, high-efficiency, and it is the most reasonable first to use RS for resource prospecting.

## 5. Conclusions

The RS prospecting model indicates mineral exploration in field and there are many new discoveries in the actual prospecting. Therefore, RS is indeed effective for resources exploration.

1) Different resources need to choose different RS information extraction. Group-anomalies extraction indicated alteration zone for metal hydrothermal mineral prospecting and electronic-anomalies was used for metal sedimentary mineral exploration.

2) It to apply By using the RS prospecting model to guide the mineral exploration, the prospecting probability can be improved. In field verification, many new mineral deposits and mineralization occurrences were discovered.

3) Important RS factors of mineral prospecting must be rare. For example the stratum contained sandstone-type copper deposits shown different color in RS lithologic enhanced image. For more accurate exploration, other information needs to overlay the RS information and analyze prospecting indicators.

**Acknowledgments**

This work was supported in part by the National Natural Science Foundation of China (No. 41102205) and Special Funds Projects for Basic Scientific Research Business Expenses of Mineral Resources Research Institutes in Chinese Academy of Geological Sciences (Nos. K1501)". We thank greatly Prof. Y.J. Zhang, Prof. J.M. Yang, X.X. Geng, J. Sun for participating research. We are grateful to relevant Bureau of Geology and Mineral Exploration and its affiliated Geological parties and related leaders and staff members for their great logistical and moral support.

**References**

[1] Kratt C, Calvin W M, Coolbaugh M F 2010 Mineral mapping in the Pyramid Lake basin: Hydrothermal alteration, chemical precipitates and geothermal energy potential *Remote Sens. Environ.* **114** 2297-304

[2] Rajendran S, Thirunavukkarasu A, Balamurugan G, Shankar K 2011 Discrimination of Iron Ore Deposits of Granulite Terrain of Southern Peninsular India using ASTER data *J. Asian Earth Sci.* **41** 99-106

[3] Gabr S, Ghulam A, Kusky T 2010 Detecting Areas of High-potential Gold Mineralization Using ASTER Data *Ore Geol. Rev.* **38** 59-69

[4] Abrams M J, Ashley R P, Brown L C, Goetz A F H and Kahle A B 1977 Mapping of hydrothermal alteration in the cuprite mining district, Nevada, using aircraft scanning images for the spectral region 0.46 to 2.36 mm *Geology* **5** 713-18

[5] Hunt G R 1979 Near-infrared (1.3-2.4μm) spectra of alteration minerals-potential for use in remote sensing *Geophysics* **44** 1974-86

[6] Tangestani M H and Moore F 2011 Comparison of three principal component analysis techniques to porphyry copper alteration mapping: a case study, Meiduk area, Kerman, Iran *Can. J. Remote Sens.* **27** 176-81

[7] Crowley J K, Hubbard B E, and Mars J C 2003 Hydrothermal alteration on the cascade stratovolcanoes: a remote sensing survey *Geological Society of America Abstracts with Programs* **35** 552

[8] Zhang Y J, Yang J M, Yao F J, Geng X X, Yu X Z 2014 *Practical Technology of Multi-Spectral Remote Sensing Prospecting information* (Beijing: Geological Publishing House) pp 1-10 (In Chinese with English Abstract)

[9] Amer R, Mezayen A E, and Hasanern M 2016 ASTER spectral analysis for alteration minerals associated with gold mineralization *Ore Geol. Rev.* **75** 239-51

[10] Ding C, Li X Q, Liu X N and Zhao L T 2015 Quartzose-mafic spectral feature space model: A methodology for extracting felsic rocks with ASTER thermal infrared radiance data *Ore Geol. Rev.* **66** 283-92

[11] Calvin W M, Littlefield E F and Kratt C 2015 Remote sensing of geothermal-related minerals for resource exploration in Nevada *Geothermics* **53** 517-26

[12] Meer F V D, Hecker C, Ruitenbeek F V, Werff H V D, Wijkerslooth C D, Wechsler C 2014 Geologic remote sensing for geothermal exploration: A review *Int. J. Appl. Earth Obs.* **33** 255-69

IOP Conf. Series: Earth and Environmental Science 74 (2017) 012015 doi:10.1088/1755-1315/74/1/012015

# Assessing the accuracy of satellite-derived urban extent over major urban clusters in the WRF model

**Ronghan XU[1], Zharong PAN[2], Hao GAO[1]**
[1] National Satellite Meteorological Center, China Meteorological Administration, Beijing 100081, China
[2] General Institute of Water Resources and Hydropower Planning and Design, MWR, Beijing 100120, China

Corresponding author: Zharong PAN (E-mail: panzharong@giwp.org.cn)

**Abstract.** Urban land use data play a central role in climate change assessments of urbanization process. The Moderate Resolution Imaging Spectroradiometer (MODIS) land use data is employed in the Weather Research and Forecasting (WRF) model. It is important to understand scaling effects of the MODIS urban data before applying to regional climate modelling. In this study, we took the Landsat derived National Land-Use/Land-Cover Dataset (NLCD) of China as the reference data to assess the accuracy of the MODIS urban map. Urban area sizes and spatial agreement of urban pixels were investigated as assessment methods at the national and metropolitan levels over China. Results showed that the accuracies vary from region to region and highlighted strengths and weaknesses of the MODIS data in different metropolitan area. The study provides insights to model communities with the suitability of the MODIS urban data for specific regional modeling.

## 1. Introduction

Land use changes play a central role in driving forces of changes in the Earth system and climate in particular [1]. The underlying surface properties change with the twin processes of urbanization and deforestation accompanied with human activities. Urban land use change, typically from vegetated surfaces to imperious built-up surfaces, can have significant impacts on meteorological elements and pollution concentrations, as it determines the exchange of heat and momentum between the land and air [2] by factors such as the surface soil moisture, albedo, and roughness length. Numerical models have been used extensively to investigate effects of direct and indirect urban land use change mechanisms on climate [3]. Previous studies have found that urban land use data with differing accuracies incorporated into regional climate modeling directly influenced the model outputs [4]. Cheng et al. [5] demonstrated that application of the inaccurate representations of land use and land cover patterns indicated overprediction or underprediction of meteorological parameters, e.g., temperature and wind speed. Therefore, it is important to validate the urban land use data in models before using them for climate studies.

Currently there are several global land use data in response to the need for analyses of the connections between natural and anthropogenic processes. For widely used community models such as the WRF model, beginning with its Version 3.1, the model provides the Moderate Resolution Imaging Spectroradiometer (MODIS) data as an alternative land use dataset. The MODIS data was acquired in 2001 at 1-km spatial resolution from the TERRA MODIS instrument [6]. As current heterogeneous maps persist in differences of urban category across geographical scales and world regions, efforts have been made to establish their accuracies. Ran et al. [7] found that the urban and water categories have the highest consistency in the area sizes but have high mixed error in China. These studies have certainly improved our understanding of accuracy and comparability between existing global urban maps. However, the validation and accurate map for regional climate model application remains inadequate [8]. Very little assessment has been undertaken to directly examine the accuracy of urban land use data especially in metropolitan regions over China,

In this study, the MODIS urban land use data in the WRF model was assessed by comparing to the finer resolution of Landsat derived urban data at the national and metropolitan levels over China. Urban area sizes and spatial agreement of urban pixels were investigated, in order to outline (i) the comparisons of accuracies of the MODIS urban data in different metropolitan areas over China and (ii) the errors of urban land use data when applying to regional modeling of urbanization effects.

## 2. Data and methods

### *2.1. Datasets and preprocessing*

Four major metropolitan areas over China were investigated since they have experienced remarkable economic growth and urban development during the past two decades, including the Beijing-Tianjin-Hebei region, the Pearl River Delta, the Yangtze River Delta, the Chengdu-Chongqing area.

The 1-km MODIS dataset derived in 2001 used by the WRF model was classified by the International Geosphere-Biosphere Programme (IGBP) modified categories. As our interest was focused on urban, the "Urban and Built-Up" category in the MODIS dataset was extracted for accuracy assessment. The extracted data was classified into urban and non-urban categories at 1-km resolution for better comparison.

The National Land-Use/Land-Cover Dataset (NLCD) of China acquired in 2000 was taken as the reference dataset. It was constructed by the Chinese Academy of Sciences from Landsat ETM (Enhanced Thematic Mapper) data [9]. The Landsat images were georeferenced and orthorectified using ground control points and high-resolution digital elevation models. Visual interpretation and digitization were applied to the Landsat data to generate thematic maps of land use and land cover. The NLCD dataset was classified by a hierarchical classification system of 25 land categories grouped into 6 aggregated classes of land types with the overall accuracy of 95%. The NLCD dataset was gridded by land use percentage data at 1-km resolution, which represents the area composition of each land use type within each one square kilometer unit. To cope with the MODIS urban and non-urban data at 1-km resolution, we used the majority-rule approach to search for the "built-up area" category in NLCD dataset with the highest frequency within the 1-km grid cell [10]. Finally, the MODIS and NLCD urban data were reprojected and registered to Albers equal area conic projection at a spatial resolution of 1-km with urban and non-urban categories.

### *2.2. Assessment methods*

In this study, urban area sizes and spatial agreement of urban pixels were investigated as assessment methods at the national and metropolitan levels. First, the comparison of urban area sizes provides a direct assessment of data accuracy in terms of the quantity. The assessment compared the area mapped as urban land in the MODIS data against in the NLCD map within the same extent. The bias of urban area sizes was calculated using the following equation.

$$O = (M - N) / N \times 100\% \tag{1}$$

In equation (1), O denotes the bias of urban area sizes of the MODIS data, M and N denote the urban area sizes of the MODIS and NLCD data, respectively.

Second, the comparison of spatial agreement considers in terms of location consistency (pixel by pixel). The MODIS and NLCD data were overlaid for comparison. The pixels with the same urban or non-urban category in both MODIS and NLCD data retained their class values (i.e. urban and non-urban) and were regarded as agreement. Whereas pixels with different categories were assigned the new class values (i.e. urban/non-urban and non-urban/urban) and considered as disagreement. Thus, there were four class values after spatial overlaying comparison, i.e., urban, urban/non-urban, non-urban/urban, and non-urban. The non-urban pixels were disregarded from the calculation of overall agreement since our attention was concentrated on the accuracy of urban category. The agreement of urban category between the MODIS and NLCD data was calculated using the following equation.

$$D = A/(A+B+C)\times 100\% \tag{2}$$

In equation (2), D is spatial agreement of urban category between the MODIS and NLCD data, A, B, and C represent the number of pixels for urban, urban/non-urban, and non-urban/urban after overlaying comparison, respectively.

## 3. Results

### *3.1. Urban area sizes*

Figure 1 illustrates the remapped urban and non-urban data derived from the MODIS against NLCD data for the four metropolitan areas. Results show that the urban land patterns for each metropolitan region between the MODIS and NLCD data are similar. Geographically, urban pixels gather in the MODIS data, whereas there are many fragile urban patches demonstrated in the NLCD data. The statistical urban area sizes both at the national and metropolitan levels are listed in Table 1.

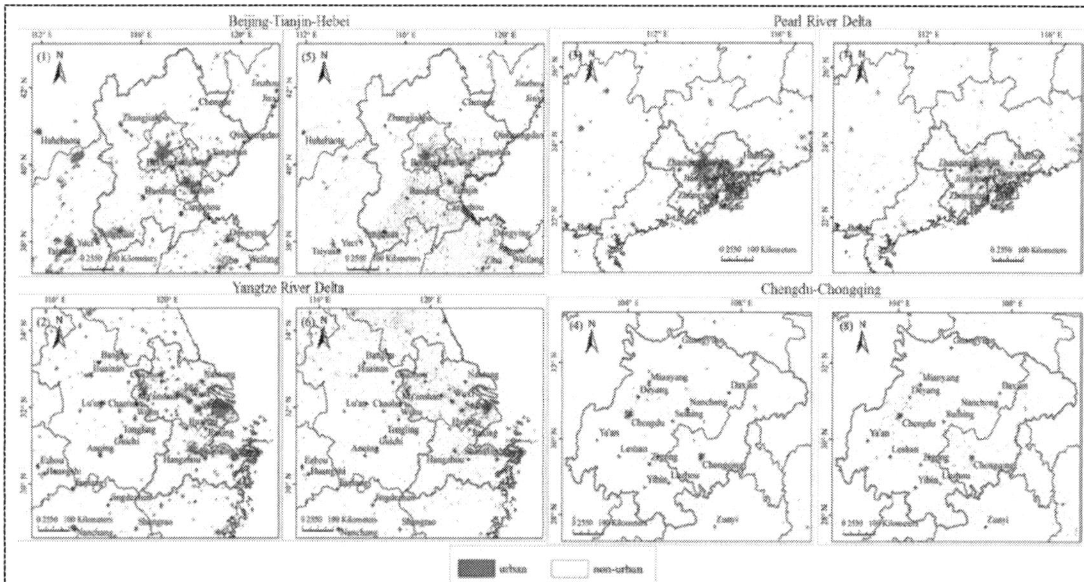


**Figure 1.** Spatial distribution of urban land remapped from the MODIS versus NLCD data in the four metropolitan areas at a 1 km resolution. (1)-(4) are remapped from the MODIS data (Left). (5)-(8) are remapped from the NLCD data (Right).

At the national scale, the total area of urban land for the whole country estimated from the MODIS and NLCD are 93444 and 64705 km2, respectively. The percentage of urban land in the NLCD data for the whole China is 0.68%. Results indicate that urban land totals predicted from the MODIS map across China has a discrepancy relative to the NLCD map. The bias from the NLCD is 44.42%. At the metropolitan scale, the greatest percentage of urban land estimated from the NLCD data is 6.17% for the Pearl River Delta. The biases of the four metropolitan regions with respect to the NLCD are all

IOP Conf. Series: Earth and Environmental Science 74 (2017) 012015 doi:10.1088/1755-1315/74/1/012015

positive. Nevertheless, the deviations of each metropolitan region are considerably different (Table 1). Results indicate that urban area sizes for the Beijing-Tianjin-Hebei and Chengdu-Chongqing region estimated from the MODIS data are consistent with the NLCD data. The bias from the NLCD data is 9.92% and 16.11% for these two metropolitan areas respectively. Whereas the Pearl River Delta (147.47%) and Yangtze River Delta (148.08%) have much higher biases relative to the NLCD data.

**Table 1.** Urban area sizes ($km^2$) and biases of the MODIS data relative to the NLCD data. The list of Percentage represents for the percentage of urban land in that item calculated from the NLCD data. The number of urban pixels for spatial overlaying comparison and agreements of distribution at the national and metropolitan levels.

| | Urban area sizes assessment | | | |
|---|---|---|---|---|
| Region | MODIS | NLCD | Percentage | Bias |
| Beijing-Tianjin-Hebei | 8288 | 7540 | 4.15% | 9.92% |
| Yangtze River Delta | 10635 | 4287 | 4.33% | 148.08% |
| Pearl River Delta | 8347 | 3373 | 6.17% | 147.47% |
| Chengdu-Chongqing | 1723 | 1484 | 0.59% | 16.11% |
| Whole China | 93444 | 64705 | 0.68% | 44.42% |
| | Spatial agreement of urban pixels | | | |
| Region | urban/urban | urban/non-urban | non-urban/urban | agreement |
| Beijing-Tianjin-Hebei | 2517 | 5771 | 5023 | 18.91% |
| Yangtze River Delta | 2676 | 7959 | 1611 | 21.85% |
| Pearl River Delta | 2468 | 5879 | 905 | 26.68% |
| Chengdu-Chongqing | 554 | 1169 | 930 | 20.88% |
| Whole China | 21217 | 72227 | 43488 | 15.49% |

### *3.2. Spatial agreement of urban pixels*

The agreements of spatial distribution of urban per-pixel comparison at the national and metropolitan levels are listed in Table 1. For the whole China, spatial agreement of the MODIS data versus the NLCD data is 15.49%. With respect to the metropolitan areas, the greatest agreement between the MODIS and NLCD data is observed in the Pearl River Delta (26.68%). The Beijing-Tianjin-Hebei region has the lowest agreement (18.91%) among the four metropolitan areas. Figure 2 illustrates spatial distributions of urban pixels of the MODIS versus NLCD data for the four metropolitan areas. Overall, pixels for the agreement (urban/urban) are surrounded by pixels for the disagreement (urban/non-urban). For the Beijing-Tianjin-Hebei region, pixels for the disagreement (non-urban/urban) are mainly distributed along the coastline. It is because there are areas used for settlements in villages and transportation mapped in the NLCD data. For other three metropolitan areas, pixels for the disagreement (non-urban/urban) are those scatted urban patches for small-medium cities and rural settlements.

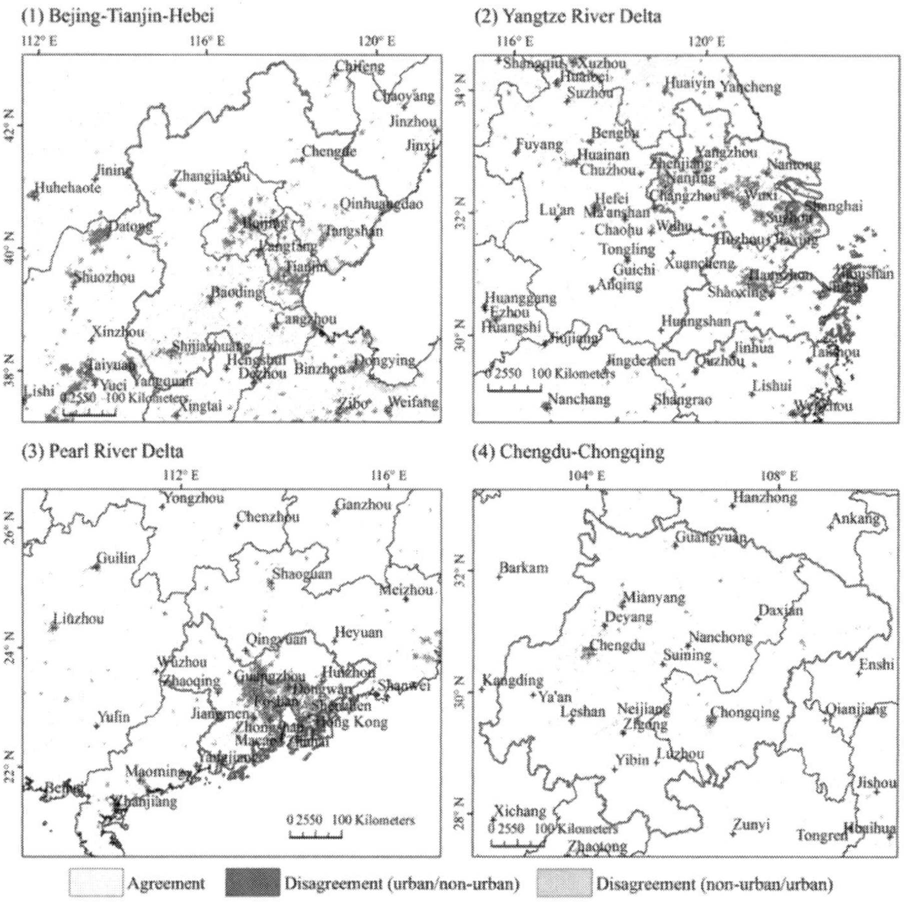


**Figure 2.** Spatial agreement of urban pixels derived from the MODIS data against the NLCD data for the four metropolitan areas: (1) the Beijing-Tianjin-Hebei region, (2) the Yangtze River Delta, (3) the Pearl River Delta, (4) the Chengdu-Chongqing region.

## 4. Discussion

This study reveals that there are varying levels of discrepancies between the MODIS urban data for the four metropolitan areas. Urban area sizes and spatial agreement of urban pixels vary from region to region. The first two assessments provide the quantity of size and pixel location. Although the size bias for the Beijing-Tianjin-Hebei area is the minimum, the spatial agreement of urban pixels for this region is the lowest. It is because the imbalance of urban distribution between the MODIS and NLCD data (Figure 1). The numbers of pixel disagreement for urban/non-urban and non-urban/urban are comparable in the Beijing-Tianjin-Hebei area (Table 1). Thus, the errors of urban fractions should be concerned as the underlying parameters when applying the MODIS data to modeling the effects of urbanization.

## 5. Conclusions

Satellite imagery plays an important role in aiding the production of urban maps for urbanization issue analysis. Users of urban maps need to be made aware of the applicability and accuracy of each map. This study outlines strengths and weaknesses of the MODIS data at the metropolitan and national levels. Assessments of urban area sizes and spatial agreement of urban pixels provide the quantity of size and pixel location. This study provides insights to model communities with the suitability of the MODIS urban data for specific regional modeling.

## 6. References

[1] Verburg P H, Neumann K, and Nol L 2011 Challenges in using land use and land cover data for global change studies *Global Change Biol* **17** 974–89

[2] Souch C, and Grimmond S 2006 Applied climatology: urban climate *Prog. Phys. Geog.* **30** 270–79

[3] Miao S, Chen F, LeMone M A 2009 An Observational and Modeling Study of Characteristics of Urban Heat Island and Boundary Layer Structures in Beijing *J. Appl. Meteorol. Clim.* **48** 484–501

[4] Oleson K W, Bonan G B, Feddema J, Vertenstein M and Grimmond C S B 2008 An urban parameterization for a global climate model. Part I: Formulation and evaluation for two cities. *J. Appl. Meteor. Climatol.* **47** 1038–60

[5] Cheng F Y, Hsu Y C and Lin P L 2013 Investigation of the effects of different land use and land cover patterns on mesoscale meteorological simulations in the Taiwan area *J. Appl. Meteor. Climatol.* **52** 570–87

[6] Friedl M A, McIver D K, Hodges J C 2002 Global land cover mapping from MODIS: algorithms and early results *Remote Sens. Environ.* **83** 287-302

[7] Ran Y H, Xin L, and Ling L 2010 Evaluation of four remote sensing based land cover products over China *Int. J. Remote Sens.* **31** 391-401

[8] Gao H, and Jia G S 2013 Assessing disagreement and tolerance of misclassification of satellite-derived land cover products used in WRF model applications *Adv. Atmos. Sci.* **30** 125–41

[9] Liu J, Tian H, Liu M 2005 China's changing landscape during the 1990s: Large - scale land transformations estimated with satellite data *Geophys. Res. Lett.* **32** L02405

[10] Wu W, Shibasaki R, Yang P 2008 Validation and comparison of 1 km global land cover products in China *Int. J. Remote Sens.* **29** 3769-85

## Acknowledgments

This study was supported by National Key Research and Development Program of China (2016YFA0600303), National Natural Science Foundation of China (41405064, 41675079).

IOP Conf. Series: Earth and Environmental Science **74** (2017) 012016 doi:10.1088/1755-1315/74/1/012016

# Rainy season change in Sanjiangyuan, China area based on the meteorological stations data

**Zhanqing Cai [1,3,4], DailiangPeng[2],Jingyi Wen [1,3], Zhi Gong [1,3,4], Tiantian Wang [1,3],Yuekai Hu [1], Yuxi Wu [1], Junfeng Xu [1,3*]**

1 Institute of Remote Sensing and Earth Sciences, Hangzhou Normal University, Hangzhou 311121,China;
2 Key Laboratory of Digital Earth Science, Institute of Remote Sensing and Digital Earth, Chinese Academy of Sciences, Beijing 100094, China;
3 Zhejiang Provincial Key Laboratory of Urban Wetlands and Regional Change, Hangzhou Normal University, Hangzhou 311121,China;
4 College of Life Environmental Sciences, Hangzhou Normal University, Hangzhou 311121,China;

Email:junfeng_xu@163.com

**Abstract**. The Sanjiangyuan area is famous as the Chinese water tower, it is a transition zone of semi-humid, semi-arid and arid area. An area of about 400,000 square kilometers and the water source of billions of people, also the ecological barrier of the whole of Asia's economic and social development.Based on the daily rainfall data of 13 meteorological stations in Sanjiangyuan area from 1985 to 2015, the rainfall indexes such as the rainy season from the beginning and the ending of the rainy season were extracted. And the trend of rainy season in the study area was analyzed. The results show that the daily rainfall in the Sanjiangyuan area accounts for more than 50% of the annual rainfall; the onset of the rainy season in the eastern part of Sanjiangyuan has a delayed trend and the central area has a tendency to advance; the number of days in the rainy season has decreased in the past 30 years, reduced 4d. There is a positive correlation between the starting date of rainy season and the average rainfall in the rainy season. The changes of these rainy seasons are of great significance to the economic development and ecological protection of the region.

## 1. Introduction

Rainy season contains a lot of important phonological information, such as the beginning and ending of the rainy season, the number of rainy days, the number of intervals, the maximum rainfall, the average rainfall [1]. These indicators have been the focus of meteorological services and meteorological research [2]. Through these indicators we can know the rainy season in a region over the years changes in the situation, while these changes will affect the region's climate, ecology, economy, people's livelihood, etc. [3-5]. The Sanjiangyuan area is a transitional zone of semi-humid, semi-arid and arid areas. It is also a typical sensitive area of the climate change. It is a fragile area of geographical environment change [6]. The impact of changes in the rainy season is much greater, will

* Corresponding author.

profoundly affect the whole of China and the global climate change. The area has important geographical and geographical advantages in the Belt and Road. It not only has rich mineral resources, water resources, or an important ecological barrier, it is the main transportation hub linking east and west, getting through the north and south and connecting the whole Asia. Therefore, the development of the region and the ecological protection work will be the focus of connecting China and the West. We use the daily rainfall data of 13 meteorological stations in the study area to extract the phonological indexes of rainy season in the past years and analyze their temporal and spatial changes, and provide scientific guidance data for the infrastructure construction and ecological protection in this area.

## 2. Materials and methods

In this study, the precipitation data of 13 meteorological stations in Sanjiangyuan area from 1985 to 2015 were used to extract and analyze phonological indexes of rainy seasons in the study area from the meteorological data service network of China(http://data.cma.cn/site/index.html). According to the Qinghai Province announced the rainy season to determine the standard C1.5 to determine the annual rainy season [7]. Which, C5, C10 and C15 are not less than 1.5 when the rainy day. The data of all the days in the whole year were collected from the rainy season, the number of rainy seasons and rainfall in the rainy season. The site in the study area distribution is shown in figure 1.The study area is divided into three parts: the east, middle and west, the east including the five sites of Banma, Jiuzhi, Henan, Xinghai and Dari. Qumalai, Qingshuihe, Yushu, Nangqian, Zaduo and other six sites are in middle; the west, including Wudaoliang and Tuotuohe .

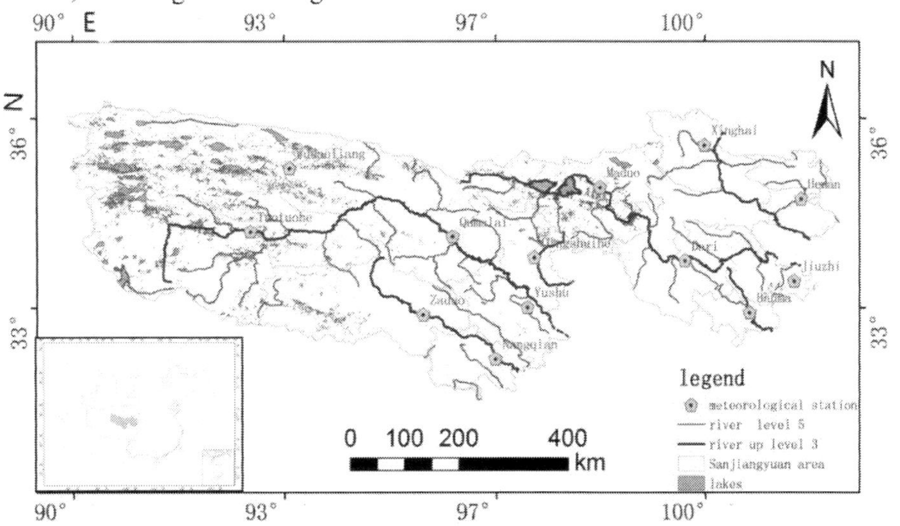


**Figure 1.** The study area distribution of meteorological stations

## 3. Results analysis

We will focus on the analysing of the annual precipitation in the rainy season and its proportion in every year; the starting and ending time of rainy season and the number of rainy days in each year; the relationship between the starting date of rainy season and the average daily rainfall in rainy season.

### *3.1. The beginning and ending of the rainy season*

Wet and dry season, the region is clearly the beginning of the rainy season is one of the main features of seasonal conversion [6]. The precipitation relative coefficient C is used to determine the relative coefficient of rainy season and non-rainy season.

$$C = \frac{R(N)}{N} / \frac{R(Y)}{Y}$$

Where R (N) is the amount of precipitation in a certain period of time, N is the number of precipitation days for 5 days, 10 days and 15 days, and R(Y) is the average annual precipitation. The precipitation relative coefficients C5, C10, C15 were calculated for 5d, 10d, and 15d, respectively. The beginning of the rainy season standards: C5, C10, C15 ⩾ 1.5 [8]. The rainy season is determined to be the beginning rainy day when it was first time satisfy the condition throughout the year,. The last rainy season day is the ending of the rainy season. Since we determine rain relative use of the rainfall coefficient, the difference between the different sites may be larger.

Using a linear equation that to charge the trend of the rainy season from the beginning and the ending.

$$Y = a_0 + a_t$$

Form the linear trend term, at> 0 means that the rainy season has a delay trend since the beginning of the rainy season, at<0 indicates that the rainy season has a early trend, at × 10 means the trend of change every 10 years (units of d / 10a)[8]. Fig.2 shows the variation trend of the heterogeneous rainy season of the Henan and Zaduo. After the judgment of the rainy season over the past thirty years, we found that,the beginning of the rainy season of 6 sites including Zaduo, Yushu, Tuotuohe, Qumalai, Qingshuihe and Nangqian had a early trend, and the early trends were 5d, 1d, 4d, 2d, 1d , 3d, and the seven sites of Banma, Dari, Henan, Jiuzhi, Maduo, Wudaoliang, Xinghai had a delay trend, and delayed 1d, 8d, 5d, 7d, 11d, 1d. As the trend of the rainy season over the years in addition to the heterogeneous of Wudaoliang, Zaduo, Dari, Henan were delayed 5d, 8d, 2d, 2devery 10a, the other stations basically delayed trend, but the trend changes in the basic 1d every 10a. The rainy season beginning day advanced in the central area mostly,and delayed in the east .

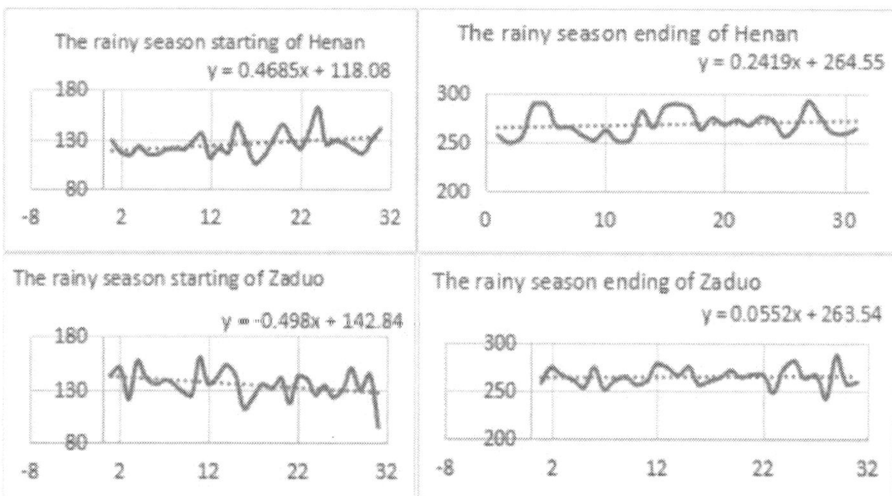


**Figure 2.** The trend of the rainy season starting and ending of Henan and Zaduo stations

### 3.2. *Rainfall in the rainy season*

Rainy season refers to the relative concentration of precipitation in a year. Only the dry and wet season in the Sanjiangyuan area, rainfall in the rainy season occupies a very high proportion in the annual precipitation [9]. Table 1 shows the annual precipitation in the rainy season, the average number of days in the rainy season and the trend of the rainy days in each year. It was found that only the rainy days (C ≧ 1.5) accounted for the lowest42% of the total precipitation in the whole year in Maduo, and the highest was 54% in Tuotuohe, with an average 50% of all. The most of rainy days in the Xinghai is 92 days. The average number of rainy days on all sites is 78 days, accounting for 21.4% of the whole year. That is, 78 days during the rainy season in the region daily rainfall accounted for half of the annual rainfall. It is found that all the 13 stations in the study area have a decreasing tendency in the past 30 years, and the decrease of the Tuotuohe is the least, and the decrease of the Wudaoliang is the most.The average of all reduced 4 days every 10a.

**Table 1.** The annual precipitation, average number of days and the trend of the rainy daysin the rainy season

| Station | Banma | Dari | Henan | Jiuzhi | Maduo | Nanhqian | Qingshuihe | Qumalai | Tuotuohe | Wudaoliang | Xinghai | Yushu | Zaduo |
|---|---|---|---|---|---|---|---|---|---|---|---|---|---|
| Rainy day rain ratio(%) | 43.21 | 43.18 | 42.77 | 43.41 | 42.01 | 49.45 | 44.29 | 50.16 | 53.99 | 44.34 | 50.89 | 48.72 | 49.89 |
| Rainy days(d) | 78 | 73 | 74 | 80 | 71 | 79 | 75 | 80 | 75 | 79 | 92 | 82 | 79 |
| Days change in rainy every 10 years（d/10a） | -7 | -6 | -3 | -4 | -3 | -5 | -5 | -2 | 0 | -9 | -4 | -4 | -1 |

### *3.3. The relationship between the beginning of rainy season and rainfall*

Many scholars have shown that the delay or early onset of rainy season will have an impact on rainfall in the rainy season [9-12]. We analyze the site rainy season starting date and the average rainy season rainfall. Table 2 summarizes the correlation (R) of the beginning of rainy season to the average rainy season for all stations. Among them, the correlation of Wudaoliang sites is 0.98, and the lowest is 0.32, the average is 0.58. Indicating that there is a positive correlation between the onset of rainy season in the study area and the average rainfall in the rainy season, that is, the rainy season will be delayed in the rainy season. This phenomenon can be based on the beginning of the rainy season this year compared with previous years, you can predict the rainy season this year, the size of the average rainfall, to guide the work carried out is significant.

**Table 2.** The correlation between the beginning of the rainy season to the average rainfall

| Station | Banma | Dari | Henan | Jiuzhi | Maduo | Nanhqian | Qingshuihe | Qumalai | Tuotuohe | Wudaoliang | Xinghai | Yushu | Zaduo |
|---|---|---|---|---|---|---|---|---|---|---|---|---|---|
| Correlation（R） | 0.57 | 0.68 | 0.49 | 0.47 | 0.68 | 0.61 | 0.32 | 0.39 | 0.38 | 0.98 | 0.81 | 0.69 | 0.53 |

## 4. Conclusion and discussione

### *4.1. Conclusions*

During the rainy season, 21.4% of the rainfall in the rainy season accounted for 50% of the annual rainfall. The beginning of the rainy season in the eastern part of Sanjiangyuan has a delayed trend and the central region has an early trend. The number of rainy days in the study area has a tendency to shorten, and it is shortened by 4 days every 10 years in recent 30 years. There is a positive correlation between the beginning rainy season and the average rainfall in the rainy season.

### *4.2. Discussion*

In this study, only the meteorological station rainfall data analysis of the study area during the rainy season, due to the less meteorological stations in the study area, to get the spatial distribution of precipitation in the study area also need to use remote sensing rainfall data.

Using a non parametric analysis method to analysis the variability in the rainy season will be an improvement.

**Acknowledgments:** This work is financially supported by the Youth Innovation Promotion Association, Chinese Academy of Sciences (Grant No. Y4YR1300QM), National Natural Science Foundation of China (Grant No. 41402304), and the Provincial Natural Science Foundation of Zhejiang (Grant No. LQ13D020002, No. LY16D010007).

**References**

[1] Wang Z, Ding Y 2008 Climatological characteristics of Chinese rainy season *Chinese Journal of Atmospheric Sciences* **32** 1-13 (In Chinese with English Abstract)
[2] Zhou S W, Jia L 1999 Climate characteristics and circulation analysis of rainy season in Tibet plateau *Meteorology* 1999 **12** 38-42 (In Chinese with English Abstract)
[3] Wang S F, Zhang H J 2011 Technical measures for road construction in rainy season *Academic Library of civil engineering* **15** (In Chinese with English Abstract)
[4] Wang B 2002 Rainy Season of the Asian - Pacific Summer Monsoon *Journal of Climate* **15** 386-98 (In Chinese with English Abstract)
[5] Wang W C, Ge Q, Hao Z 2008 Rainy Season at Beijing and Shanghai since 1736 *J. Meteorol. Soc. Jpn.* **86** 827-34
[6] Dai J H, Wang M M, Wang H H 2009 Climate change and ecological effect of the semi humid and semi arid transitional zone in the eastern part of Northwest China *Quaternary Res.* **29** 920-30
[7] Zhao J F, Guo J P 2010 Trend of wetting and drying in China *Journal of agricultural engineering* **26** 18-24 (In Chinese with English abstract)
[8] Huang X Q, Yang Y 2008 Climate change trend in the beginning of rainy season in Tibet plateau *Tibet science and technology* **3** 64-65
[9] Liu Y Y, Zhang X Q, Sun Y 2011 Temporal and spatial variation characteristics of precipitation in the rainy season in the arid region of Northwest China under the background of global warming *Advances in climate change* **07** 97-103 (In Chinese with English Abstract)
[10] Wu H Q 2003 Analysis of the distribution of rainfall patterns in rainy seasons beginning in Guangxi and in the flood season Tropical geography **23** 126-29 (In Chinese with English Abstract)
[11] Feng Z Q 1987 Characteristics of precipitation and rainy season in northwest arid area *Arid area agricultural research* **2** 21-31
[12] Liu Y Analysis of the climatic characteristics of the late rainy season in Yunnan *Meteorology* **26** 45-49 (In Chinese with English Abstract)
[13] Wang Y, Lu X L 2005 Regional characteristics of rainy season precipitation in eastern China *Journal of atmospheric science* 28 609-16 (In Chinese with English Abstract)

IOP Conf. Series: Earth and Environmental Science **74** (2017) 012017 doi:10.1088/1755-1315/74/1/012017

# Research on trend of warm-humid climate in Central Asia

**Zhi Gong [1,3,4], Dailiang Peng [2], Jingyi Wen [1,3], Zhanqing Cai [1,3,4], Tiantian Wang [1,3,4],Yuekai Hu [1], Yaxin Ma [1], Junfeng Xu [1,3*]**

1 Institute of Remote Sensing and Earth Sciences, Hangzhou Normal University, Hangzhou 311121,China;
2 Key Laboratory of Digital Earth Science, Institute of Remote Sensing and Digital Earth, Chinese Academy of Sciences, Beijing 100094, China;
3 Zhejiang Provincial Key Laboratory of Urban Wetlands and Regional Change, Hangzhou Normal University, Hangzhou 311121,China;
4 College of Life Environmental Sciences, Hangzhou Normal University, Hangzhou 311121,China

*Email: junfeng_xu@163.com
**Abstract**. Central Asia is a typical arid area, which is sensitive and vulnerable part of climate changes, at the same time, Central Asia is the Silk Road Economic Belt of the core district, the warm-humid climate change will affect the production and economic development of neighboring countries. The average annual precipitation, average anneal temperature and evapotranspiration are the important indexes to weigh the climate change. In this paper, the annual precipitation, annual average temperature and evapotranspiration data of every pixel point in Central Asia are analyzed by using long-time series remote sensing data to analyze the trend of warm and humid conditions. Finally, using the model to analyzed the distribution of warm-dry trend, the warm-wet trend, the cold-dry trend and the cold-wet trend in Central Asia and Xinjiang area. The results showed that most of the regions of Central Asia were warm-humid and warm-dry trends, but only a small number of regions showed warm-dry and cold-dry trends. It is of great significance to study the climatic change discipline and guarantee the ecological safety and improve the ability to cope with climate change in the region. It also provide scientific basis for the formulation of regional climate change program. The first section in your paper

## 1. Introduction:

Central Asia is the largest main non-zonal arid regions in the world. It is very sensitive to climate change, researches show that the rate of global warming in Central Asia is more than doubled in recent centuries [13],which making a series of climate security risks in Central Asia, such as poverty increasing, immigrant increasing and cross-border resource conflicts caused by climate change. At the same time, Central Asia is the key area for the implementation of the strategy "One Belt and One Road". In order to ensure the smooth implementation of this strategic initiative, it is necessary to study the climate change trend of warm and wet climate in Central Asia.

The traditional method of studying the climate change of the warm-humid climate is to use the surface meteorological station data to interpolate and obtain the climatic change status of the whole study area. However, the meteorological stations in Central Asia are sparsely distributed and unevenly

[*] Corresponding author.

distributed, the terrain span is large and the underlying surface situation is complex, the traditional methods there is a serious point to surface phenomenon, it is difficult to accurately describe the region's climate trends and the spatiotemporal pattern distribution [5].

The feature of remote sensing data is spatial continuity, which can overcome the limitation of the sparseness site data and the inherent error of spatial interpolation data. It has become an important data source of climate change research in the region (especially in Central Asia, where the climate data is scarce and the terrain is complex) .The study of warm-humid climate trends in Central Asia based on remote sensing data can obtain the rich, comprehensive and scientific understanding of the "Silk Road Economic Belt". Thus it also provide scientific advisory and support for the construction of the "Silk Road Economic Belt".

In this paper, the trend of annual mean temperature, precipitation and evapotranspiration in Central Asia is analyzed by using multiple remote sensing data, and using the model to study the trend of warm-humid climate in Central Asia. It is of great significance to respond to climate change to make the appropriate decision-making.

## 2. The data sources and methods of the warm-humid climate trend

In this paper, the region of Central Asia, including the five Central Asian countries and China's Sinkiang, the latitude and longitude range of 50 ° -90 ° E, 35 ° -55 ° N, which is defined as the study area. The southeastern terrain is lower than the northwestern part, and it is the typical temperate desert steppe continental climate, where the annual precipitation in 100 ~ 400mm, the Mountain is slightly higher than others, however the height of the desert areas was in 50mm below [4].

The paper mainly uses the NASA's high-resolution evapotranspiration datamod16, the temperature data mod11c3 products and the trmm satellite data. Evapotranspiration and temperature data are monthly collected values. The spatial resolution of original data is 1 km, after resample, we obtain the data that the spatial resolution is5 km. The trmm data is used on a monthly scale with a spatial resolution of 0.25 degrees for the data product.

In this study, the average annual temperature, precipitation and the annual evapotranspiration of each pixel point in China's Sinkiang and the five central Asian countries were analyzed by using a linear trend analysis method. Then obtain the annual mean temperature trends (temperature_K ), Annual precipitation trends (evapotranspiration_K) and rainfall trends (evapotranspiration_K). Finally, using the method which was named decision tree to establish the warm-humid trend monitoring model on temperature, evapotranspiration and precipitation data, last access to Central Asia warm-wet climate change trend.

## 3. Temporal and spatial variation of warm-humid climate and conclusions

Temperature, rainfall and evapotranspiration are the important indicators of warm-humid climate. In this study, we analyzed the climatic characteristics of Central Asia using by annual precipitation, annual mean temperature and annual evapotranspiration in Central Asia, so as to have a comprehensive understanding of the trend of warm and humidification in Central Asia.

### *3.1 The analysis of temporal and spatial variation on temperature*

Rapid warming of the global climate is already an indisputable fact. Central Asia is the world's largest non-zonal arid zone, which is the most sensitive to global warming. In recent 30 years, with the global warming in Central Asia, the temperature increased significantly and the rate of temperature is higher than the global average data, which makes the glaciers melt rapidly and the freshwater reserves of the glaciers decrease sharply. This is a great challenge to the fragile ecological environment of arid and semi-arid in Central Asia [5].

Although the general trend of global temperature change is almost the same, the variation trend of temperature in different regions is unequal because of sundry causes which are latitudes, terrain and land-sea distribution. In order to understand the long-term trend of temperature in Central Asia, we analyzed the trend of pixel-by-pixel variation of the annual mean temperature from 2000 to 2015 in Central Asia and China's Sinkiang based on the MODIS and the MOD11C3 products from 2000 to 2015,

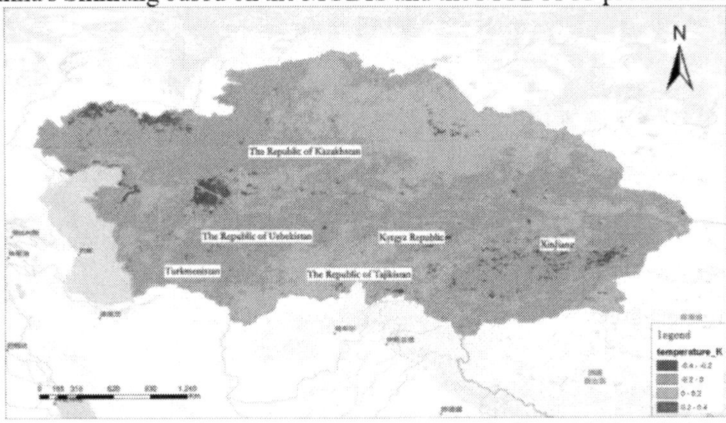


**Figure 1.** Annual average temperature trends from 2000 to 2014.

The results As shown in Fig. 1, the regression coefficients of annual mean temperature trends in Central Asia are between -0.2 and 0.2, with the north and central regions of Kazakhstan showing a trend of increasing temperature. But the west and southeast parts show a decreasing trend; The trend of temperature increasing in Turkmenistan, Uzbekistan, Kyrgyzstan, the most area of junction in Tajikistan; The border region and the central region of China's Sinkiang showed the trend of temperature increasing.

*3.2 The analysis of temporal and spatial variation on precipitation*

Rapid Precipitation in Central Asia is characterized by an increasing trend of both sides while decreasing trend of middle sides in space. High-value centers of precipitation appear in Kyrgyzstan, Tajikistan. Western part of Uzbekistan, southern part of Kazakhstan and northern part of Turkmenistan and China's Sinkiang are the low value of precipitation centers, especially in Xinjiang, China's Tarim Basin, the average annual rainfall is insufficient [2].

In order to understand the trend of annual precipitation change in Central Asia, we used the trmm satellite data 3B43 month data to analyze the annual precipitation in China.

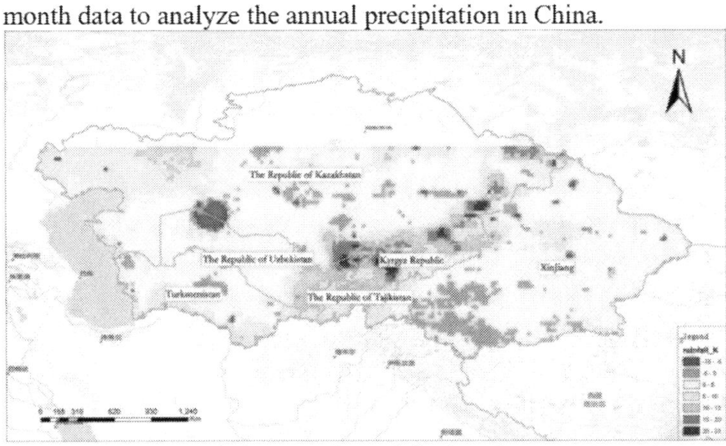


**Figure 2.** Trends in annual precipitation from 2000 to 2014

Monitoring results shown in Fig 2, trmm precipitation data cover the range of 50 ° S-50 ° N, so the data of northern region of Kazakhstan would be missing. The regression coefficients in the Central Asia region are in the range of -5 to 10, with the positive data region being the dominant region. The annual precipitation in most parts of Central Asia is increasing, with the increase in the eastern region of Tajikistan, Kyrgyzstan and Kazakhstan. The precipitation in Kazakhstan, Turkmenistan and southwestern part of China's Sinkiang is slightly decreasing.

*3.3 Analysis of temporal and spatial variation on evapotranspiration*

The arid region of Central Asia is a region where has serious shortage of water resources and very large evaporation [1]. The change of annual evapotranspiration seriously affects the humid climate in Central Asia. In order to understand the trend of annual evapotranspiration in Central Asia, we used MODIS data mod16 which spatial resolution of 1 km of annual evapotranspiration data to monitor the trend of each pixel in Central Asia.

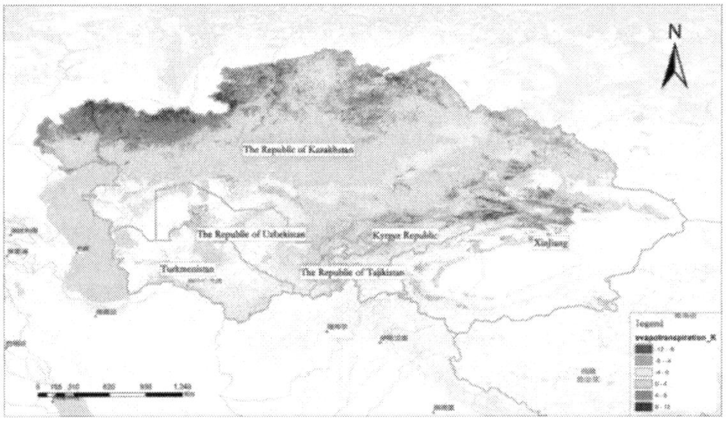


**Figure 3.** Trends of annual evapotranspiration from 2000 to 2014

Monitoring results As shown in Fig3, mod16 evapotranspiration data products are severely missing due to the low vegetation coverage in the northwestern parts of Turkmenistan and Uzbekistan and the southwest of China's Sinkiang. The annual evapotranspiration in most parts of Kazakhstan showed an increasing trend, especially in the northern region; The annual evapotranspiration in the northwestern Uzbekistan and Turkmenistan showed an increasing trend, while the southeast part showed a decreasing trend; Kyrgyzstan showed an increasing trend, but Tajikistan showed a decreasing trend; The north of Xinjiang Uygur Autonomous Region showed an increasing trend of increasing and a declining trend in the west.

*3.4 Spatial Analysis of Climate Change Trends*

The warm-humid climate is the result of the interaction of annual precipitation, annual evapotranspiration and annual mean temperature. In this study, the method of dryness index —— formula (1) , which is used to reflect the dry and wet change in Central Asia.

$$R = E / P \quad (1)$$

In the formula: r is the drought index, E is the annual evaporation, mm; P is the annual precipitation, mm;

When $r > 1$, the annual evaporation is greater than the annual precipitation, the climate is dry, as the r value larger, reflecting the climate is more dry; when $r < 1$, the annual precipitation is greater than the annual evaporation, the climate is wet, as the r value smaller, reflecting the climate is more wet.

To obtain weather change trend(r_K) from 2000—2014,we analysis the drought index-r by unitary linearity regression, then combine with temperature_K, to Draw a scatter diagram to show the trend of warm and wet change in Central Asian.

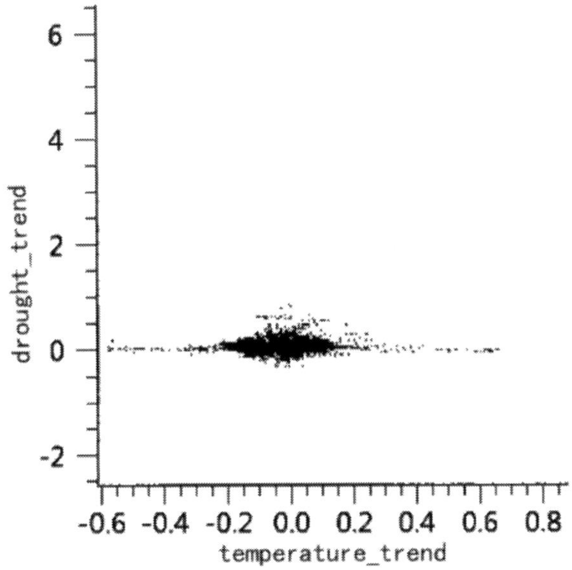


**Figure 4.** Scatter diagram of warm-humid climate change

Results as shown in figure 4 (for temperature_K abscissa and ordinate r_K - 1), on the whole area in central Asia, the change of humidity is bigger, either warm-wet or cold-wet, temperature changing trend is lesser, cold-wet, dry-wet, warm-wet and cold-wet has little difference.

Using the trend of drought index (r_K), combining with the annual temperature trend (temperature_K) to establish a decision tree model (Figure 5) to obtain the spatial distribution of warm-humid climate change trends in Central Asia.

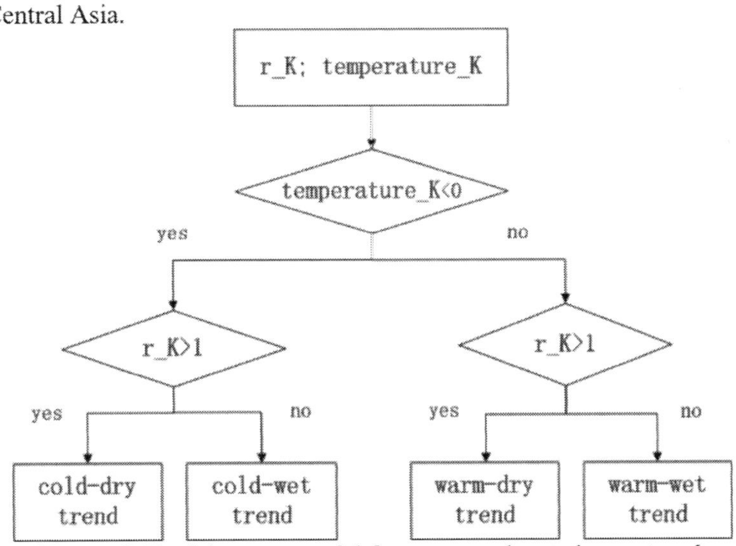


**Figure 5.** Decision tree model for warm and wet change trends

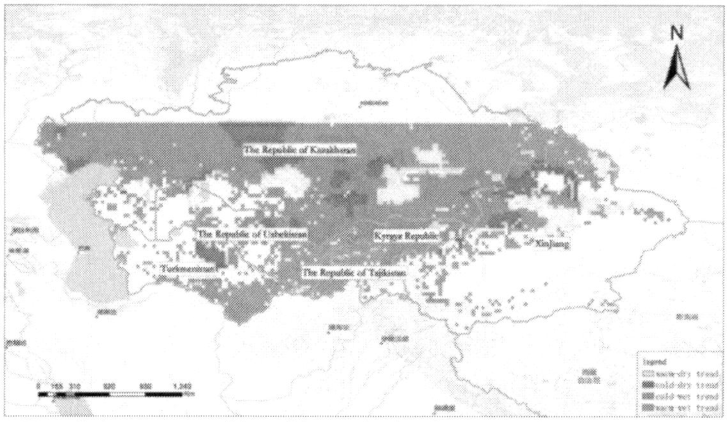


**Figure 6.** Trends in warm and humid trends for the Central Asian region, 2000-2014

The results are shown in Figure 5, because of the westerlies and global warming, the western and eastern regions of Kazakhstan were respectively showed cold-wet and warm-wet trends, a small part of the central region showed a dry-cold trend; the southern part of Uzbekistan and Turkmenistan showed warm-humid trend, but appeared cold trends in the middle part ;Tajikistan is characterized by a warm-humid trend; the northern part of Kyrgyzstan is shown cold -wet trend, but the southern showed warm-humid trend. The Xinjiang Uygur Autonomous Region has a warming trend at the junction with Kyrgyzstan and Tajikistan.

## 4. Expectation

Central Asian climate is obviously influenced by westerly circulation which with seasonal characteristics and the North Atlantic Oscillation [2]. Therefore, in order to reveal the climatic change of the warm-humid climate in Central Asia more comprehensively and meticulously, we need to use the temporal resolution of monthly scale to analyze the tendency of warm-wet climate in Central Asia by using remote sensing data.

There is lack of data on the rainfall and evapotranspiration used in the study. We need to find other data resources to replace or build up the model to invert the data of rainfall and evaporation to complement the research.

**Acknowledgments:** This work is financially supported by the Youth Innovation Promotion Association, Chinese Academy of Sciences (Grant No. Y4YR1300QM), National Natural Science Foundation of China (Grant No. 41402304), and the Provincial Natural Science Foundation of Zhejiang (Grant No. LQ13D020002, No. LY16D010007).

## References

[1] Li X W, Zhang L, Guo H D, Fu W X, Lu L L, Qiu Y B, Wang X Y, Jia G S 2016 Recognition of Eco-environment Global Change Response of Arid and Semi-arid Region of the Silk Road Economic Belt *Bulletin of Chinese Academy of Science* (In Chinese with English abstract)

[2] Wu H M, Huang A N, He Q, Zhao Y 2013 Projection of the spatial and temporal variation characteristics of precipitation over Central Asia of 10 CMIP5 models in the next 50 years Arid Land Geography, 36 669-79 (In Chinese with English abstract)

[3] Huang Q X, Zhao Y, He Q 2013 Climatic Characteristics in Central Asia Based on CRU Data *Arid Zone Research* **30** 396-403 (In Chinese with English abstract)

[4] Bromwich D H, Fogt R L, Hodges K I and Walsh J E 2007A tropospheric assessment of the ERA-40, NCEP, and JRA-25 global reanalyses in the polar regions *J. Geophys. Res.* **112** 185-194

[5] Lioubimtseva E, Cole R, Adams J M, Kapustin G 2005 Impacts of climate and land-cover changes in arid lands of Central Asia *J. Arid Environ.* **62** 285-308

[6] Huang Q X, Zhao Y, He Q 2013 Climatic Characteristics in Central Asia Based on CRU Data Arid Zone Research 30 396-403 (In Chinese with English abstract)

[7] Beijing, Beijing 2013 A study on precipitation trend and fluctuation mechanism in northwestern China over the past 60 years *Acta Physica Sinica* **62** 129201-10 (In Chinese with English abstract)

[8] Yao J Q, Liu Z H, Yang Q, Liu Y, Li C Z, Hu W F 2014 Temperature variability and its possible causes in the typicalbasinsof the arid Central Asia in recent 130 years*Acta Geographica Sinica* **69** 291-302 (In Chinese with English abstract)

[9] Yao J Q, Yang Q, Mao W Y, Xu X B, Liu Z H 2016 Evaluation of the impacts of climate change and human activities on the hydrological environment in Central Asia *Journal of Glaciology &Geocryology* **38** (In Chinese with English abstract)

[10] Shen W F, Miou Q L, Wei T X, Kong C C 2013 Analysis of temperature variation in recent 130 years in Central Asia *Journal of Arid Meteorology* **31** (In Chinese with English abstract)

[11] Sun H B, Wang R H, Yang G S 2007 The Mountain-Oasis-Desert System and Characteristics of Climate in the Arid Zone of Center Asia-A Case Study in the Northern of Xinjiang in China and the East of Kazakhstan *Journal of Arid Land Resources & Environment* **21** 6-11 (In Chinese with English abstract)

[12] Song L C, Han Y X, Sun G W 2003 The Arid Climate Variable Characteristics and Its Influence on Production Structure in the Central Asia and Northwest of China *Arid Meteorology* **21** (In Chinese with English abstract)

[13] Chen F, Wang J, Jin L, Zhang Q, Jing L and Chen J H 2009 Rapid warming in mid-latitude central Asia for the past 100 years *Front. Earth Sci.* **3** 42-50

IOP Conf. Series: Earth and Environmental Science **74** (2017) 012018 doi:10.1088/1755-1315/74/1/012018

# Analysis of Human Disturbance and Ecological Security Evolution in Oasis in Arid Area Based on LUCC: A Case Study of Oasis in the Northern Tianshan Mountain Slope Economic Zone

**W J Song[1,2], M H Chen[1*], Q Zhang[2], S S Liu[2] and J N Yang[2]**

[1] College of Geography and Environmental Sciences, Zhejiang Normal University, Jinhua Zhejiang 321004, China

[2] Institute of Remote Sensing and Digital Earth, Chinese Academy of Sciences, Beijing 100094, China

E-mail: moonchun@126.com

**Abstract.** Oases in arid areas are environmentally and economically vulnerable regions. Study on ecological security of oases in arid areas is of great significance to the stability and the economic development of oases. Based on Land Use/Land Cover data in 1965, 1980, 1995, 2005 and 2015, the study analyze the temporal and spatial changes in human disturbance and ecological security of oases in the Northern Tianshan Mountain Slope Economic Zone (NTMSEZ) in recent 50 years by establishing the ecological security index (ESI) through human disturbance index and landscape vulnerability index. The results showed that: in recent 50 years, the human disturbance of the NTMSEZ has been increased to current moderate human impacts. Urban construction, oasis expansion and farmland reclamation are the main factors of the increment. The human disturbance in Urumchi, Shihezi, Kuitun, Miquan and Changji is higher than that in other oases and that in core areas of oasis is higher than other areas. The ESI of the NTMSEZ increases firstly and then decreases. In most areas, the ESI is "relatively unsafe" and "critical". However, there are increasingly more vulnerable areas, moving northwestwards and expanding southwards. The ESI gradually presents a "NW-SE" trend of zonal distribution pattern.

* Corresponding author. Chen M H, 1976, female, Doctor, E-mail: moonchun@126.com

## 1. Introduction

With increasingly frequent global climate changes and human activities, eco-environmental issues have been a common concern [1-2]. In China, the arid areas have abundant land, light, heat and mineral resources. However, due to an arid climate, severe drought and water resources shortage, they are also the most eco-environmentally vulnerable areas in China [3]. With oasis expansion and economy development, the ecological balance has been gradually disrupted in arid areas, causing a number of eco-environmental issues. In this case, the ecological security of oases has been a research focus in recent years [4-5]. Ecological security refers to a state where there is no threat to people's daily life, health, safety, basic rights, living security sources, necessary resources, social order, ability to adapt to environmental changes, etc [6]. In recent years, from the landscape ecology perspective, the ecological security assessment system built based on the landscape index and the statistical data in general. However, when the high correlation between the selected landscape index and the human disturbance degree is not indicated, it might cause considerable errors in the results. With the statistical data as an index, there will be a loss in spatial information [7]. In this paper, oases in the Northern Tianshan Mountain Slope Economic Zone (NTMSEZ) is selected as a case study and the land cover type maps in 1965, 1980, 1995, 2005 and 2015 are taken as the main data source to establish the human disturbance degree and the ecological security index based on GIS technique. The aim of this paper is to analyze the temporal and spatial changes in human disturbance and ecological security in the area.

## 2. Data and study method

### *2.1. Study area*

The NTMSEZ is located in the middle section of northern slope of Tianshan Mountains in Xinjiang, is a typical "mountain-basin system" in northwest arid area in China. It covers Urumchi, Changji, Miquan, Fukang, Hutubi, Manas, Shihezi, Shawan, Wusu, Kuytun and Karamay. It is the most densely populated area and the most economically developed area in Xinjiang, a key area of the western development strategy and a central area.

### *2.2. Data*

For this study, the Land Use/Land Cover (LULC) data with a spatial resolution of 30m in 1965, 1980, 1995, 2005 and 2015 from Resources and Environmental Sciences Data Center, Chinese Academy of Sciences (http://www.resdc.cn) were selected as the data sources. Considering the practical situation of the study area, secondary land cover types of LULC data can be divided into 13 classes, including glaciers, shrubbery lands, beaches and flats, unused lands, sparsely forested woodland, medium/high coverage grasslands, lakes, sparsely coverage grasslands, irrigated land, reservoir and pond surface, dry land, residential quarters in rural areas, areas of cities and town.

### *2.3. Study method*

In this paper, human disturbance index (U) and landscape vulnerability index (Q) were used to establish ecological security index (ESI). Based on the different proportion of landscape types area of a grid, this paper evaluate the disturbance intensities of human activity. The calculation method can be found in the reference [8].

$$U = \sum_{i=0}^{n} (A_i / A \times P_i) \quad (1)$$

where: $U$ is human disturbance index of a grid; $A$ is the total area (km$^2$) of the grid; $A_i$ is the area (km$^2$) of the land cover type $i$ in the grid; and $P_i$ is the human disturbance class of the land cover type $i$. Following the studies of Walz [9] and Chen [10], a seven-point scale was used to classify different land cover types by degree of human disturbance (table 1).

**Table 1**. Different land cover types to degree of human disturbance.

| Class | Degree of human disturbance | Land cover type |
|---|---|---|
| 1 | almost no human impacts | glaciers, |
| 2 | weak human impacts | shrubbery lands, beaches and flats, unused lands |
| 3 | moderate human impacts | sparsely forested woodland, medium/high coverage grasslands , lakes |
| 4 | moderate-strong human impacts | sparsely coverage grasslands, irrigated land, reservoir and pond surface |
| 5 | strong human impacts | dry land |
| 6 | very strong human impacts | residential quarters in rural areas |
| 7 | excessively strong human impacts | areas of cities and town |

Landscape vulnerability (Q) refers to the vulnerability of the ecosystem represented by the land cover types when suffered external disturbances. In general, the simpler the structure of the food chain, the less the species diversity index, the more fragile the ecosystem. A ten-point scale was used to classify different land cover types by degree of vulnerability (table 2).

**Table 2**. Different land cover types to degree of vulnerability.

| Degree of vulnerability | Land cover type | Degree of vulnerability | Land cover type |
|---|---|---|---|
| 1 | areas of cities and town, residential quarters in rural areas | 6 | irrigated land |
| 2 | shrubbery lands | 7 | irrigated land |
| 3 | sparsely forested woodland, medium/high coverage grasslands | 8 | lakes |
| 4 | sparsely coverage grasslands | 9 | unused lands |
| 5 | dry land | 10 | glaciers, |

Accordingly, the ESI is introduced, which indicates the degree of ecological security of land cover type. The formula is given as follows:

$$ESI = 1 - U \times Q/70 \quad (2)$$

## 3. Analysis of results

### *3.1. Temporal and spatial changes in human disturbance degree*

In recent 50 years, human disturbance of the NTMSEZ has been increased from 2.56 in 1965 to 3.19 (moderate human impacts) in 2015. Human disturbance is divided into 6 classes to calculate the area of each class in different years (table 3). A spatial distribution map is given in figure1. It can be learnt that in recent 50 years, human disturbance of the NTMSEZ mainly falls into Class 2～3, while Class 5～6 has the most obvious changes in area, followed by Class 4～5; and the area of Class 1～2, 2～3 and 3～4 tends to decrease with fluctuated changes, while that of Class 4～5, 5～6 and 6～7 presents a significant increase trend. From the perspective of time, there were the most obvious changes in human disturbance during 1965-1980 and 2005-2015. In the NTMSEZ, there is the highest human disturbance degree along the "NW-SE" line, with gradual decrease towards both sides. With Urumchi as the boundary, both the human disturbance changes and the disturbance degree of oasis in the west are higher than those in the east. Human disturbance in each core oasis area is 5～7, higher than that in other areas. Moreover, in recent 50 years, the human disturbance has reached very strong and excessively strong human impacts. By 2015, human disturbance degree in Urumchi, Shihezi, Kuytun, Miquan and Changji has been higher than that in other oases. With increasingly intensive human activities, the areas of different human disturbance classes now have expanded from fragmentation towards integration.

**Table 3**. Statistics of area of human disturbance class in different times (km$^2$).

| Human disturbance class | 1965 | 1980 | 1995 | 2005 | 2015 | Changes in area in recent 50 years |
|---|---|---|---|---|---|---|
| **1～2** | 33316.20 | 14491.17 | 14204.79 | 13795.92 | 18813.78 | -14502.40 |
| **2～3** | 55967.58 | 50261.94 | 50478.93 | 50019.12 | 44029.71 | -11937.90 |
| **3～4** | 30502.62 | 38752.74 | 38789.64 | 38029.32 | 26447.31 | -4055.31 |
| **4～5** | 3326.31 | 19097.28 | 19101.33 | 20414.79 | 29788.74 | 26462.43 |
| **5～6** | 1.80 | 511.29 | 539.73 | 848.25 | 3551.13 | 3549.33 |
| **6～7** | 0.00 | 0.09 | 0.09 | 7.11 | 483.84 | 483.84 |

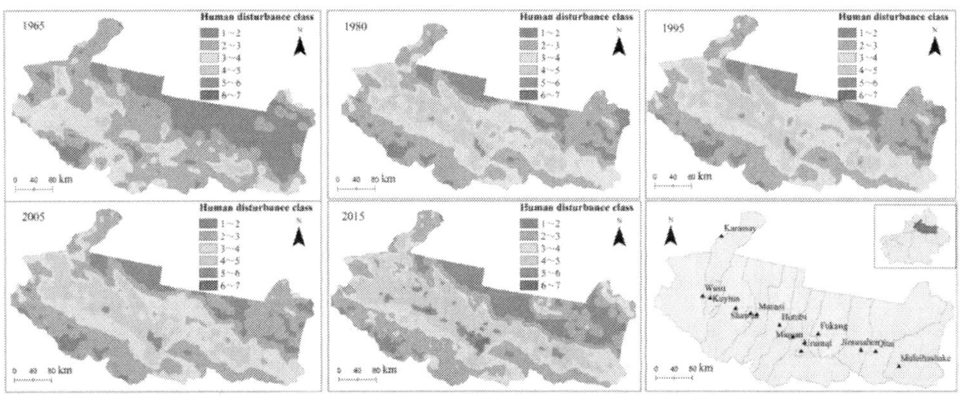


**Figure 1.** Spatial distribution of human disturbance of the NTMSEZ.

*3.2. Analysis of temporal and spatial changes in ecological security*

The ESI of the NTMSEZ in different years was calculated (table 4) as per Formula (2). The ecological security of the area under study was divided into 7 levels based on natural breaks, as shown in table 5.

**Table 4**. ESI in different years.

| | 1965 | 1980 | 1995 | 2005 | 2015 |
|---|---|---|---|---|---|
| **ESI** | 0.75 | 0.76 | 0.77 | 0.77 | 0.76 |

**Table 5**. ESI level criterion.

| Level | 1 Level | 2 Level | 3 Level | 4 Level | 5 Level | 6 Level | 7 Level |
|---|---|---|---|---|---|---|---|
| ESI | <0.65 | 0.65~0.70 | 0.70~0.75 | 0.75~0.80 | 0.80~0.85 | 0.85~0.90 | 0.90~0.95 |
| Criterion | Extremely unsafe | Unsafe | Relatively unsafe | Critical | Relatively safe | Safe | Ideally safe |

*3.2.1. Time difference.*

Statistics of areal percentage of ESI levels in different years are given in table 6. According to table 6, in recent 50 years, the mean values of the economic belt on NTMSEZ is between 0.75 and 0.8, presenting a "first increase and then decrease" trend. However, the ecological security value in 2015 is still higher than the initial value. As a whole, the ESI in most zones in the study area is "critical" and only a few zones fall into "extremely unsafe" and "unsafe". The area of Level 3 and Level 4 decreases and that of other levels increases. This is due to land reclamation on a large scale before 2005 with the purpose of meeting local residents' production and living demands and solving the population-land contradictions. In this case, the land cover type tend to be more abundant; the structures tend to be stable and the rises. After 2005, with further oasis expansion, urban construction and land reclamation, the ESI drops down.

**Table 6**. Statistics of areal percentage of ESI level in different periods.

| Level | 1965/% | 1980/% | 1995/% | 2005/% | 2015/% | Changes in area during 1965-2015 | |
|---|---|---|---|---|---|---|---|
| | | | | | | Percentage/% | Area/km$^2$ |
| 1 | 0.81 | 0.12 | 0.11 | 0.10 | 1.51 | 0.70 | 859.77 |
| 2 | 6.09 | 6.32 | 6.28 | 9.19 | 15.18 | 9.09 | 11201.40 |
| 3 | 53.74 | 46.03 | 46.08 | 44.96 | 40.16 | -13.58 | -16722.09 |
| 4 | 27.40 | 21.71 | 21.68 | 19.68 | 22.03 | -5.37 | -6575.67 |
| 5 | 7.62 | 13.82 | 13.84 | 13.45 | 8.59 | 0.97 | 1192.26 |
| 6 | 4.01 | 11.35 | 11.36 | 11.96 | 12.10 | 8.09 | 9955.26 |
| 7 | 0.36 | 0.65 | 0.65 | 0.65 | 0.43 | 0.07 | 88.65 |

*3.2.2. Spatial difference.*

According to the spatial distribution of ESI shown in figure 2, there have been increasingly more zones with risk in the NTMSEZ and the ecosystem shows the trend of instability in recent 50 years. From the spatial distribution, the ESI in the west is lower than that in the east, and that in the north is lower than that in the south. The zones of Level 1 and 2 move northwestwards and expand southwards in general. The greatest changes in the ESI are observed in northwest oases. In 1965, Level 3 and 4 predominated in the said zone. However, Level 1 and 2 was observed in most zones in 2015, indicating that the ESI was obviously decreased. The zone with the lowest ESI (Level 1) has moved from Qitai County to Urumchi and Changji and finally to Karamay, Shawan and Manas. The zones of different levels now have expanded tending from fragmentation towards integration. In 2015, Level 6 and 7 of oases in the economic belt on NTMSEZ had the largest area, mainly found along the "NW-SE" axis, i.e. the ESI in core oasis area is lower than that in other areas. The ESI has obvious spatial differentiation characteristics, presenting a "NW-SE" trend of zonal distribution pattern as a whole, lying in the Relatively Unsafe Zones, Extremely Unsafe Zones, Unsafe Zones, Safe and Relatively Safe Zones and Critical Zones from north to south respectively.

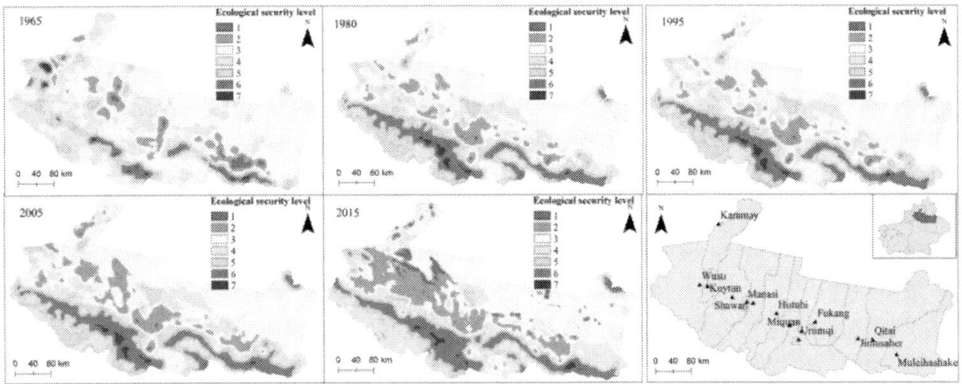


**Figure 2**. Spatial distribution of ESI of the NTMSEZ.

## 4. Conclusions

The study analyzed the temporal and spatial changes in human disturbance and ecological security of oases in the NTMSEZ in recent 50 years. The following conclusions can be drawn:

(1) From 1965 to 2015, human disturbance of the NTMSEZ was increased to a medium degree. Urban construction, oasis expansion and farmland reclamation are the main factors of the increasing trends. In 2015, human disturbance degree in Urumchi, Shihezi, Kuytun, Miquan and Changji was higher than that in other oases and that of core oasis area is higher than that in other areas.

(2) From 1965 to 2015, the ESI of the NTMSEZ increases firstly and then decreases. In most areas, the ESI is "relatively unsafe" and "critical". However, there are increasingly more areas in risk, moving northwestwards and expanding southwards. The ESI gradually presents a "NW-SE" trend of zonal distribution pattern.

## References

[1] Fan Z M, Li J and Yue T X 2013 Spatial-temporal Change of Land-Cover in Ecosystem Transitional Zones on the Loess Plateau of China *Journal of Natural Resources* **28** 426-35 (In Chinese with English Abstract)

[2] Chen X, Zhou C H 2005 Review of the Studies on Ecological Security *Progress In Geography* **24** 8-20 (In Chinese with English Abstract)

[3] Zhou J 2010 Study on urban scale of Shiyang River Basin Based on Ecological Carrying Capacity—Taking Liangzhou district as an example *Guangdong Agricultural Sciences* **5** 162-63 (In Chinese)

[4] Wang B C 2007 A Study on Change of the Ebinur Lake Oasis in Recent Thirty Years Based on Digital RS Images *Bulletin of Soil and Water Conservation 2* 107-11 (In Chinese with English Abstract)

[5] Ma K, Wang Z, Wu Y J, Zhou M Y and Shi Q D 2011 Analysis of the Landscape

Eco-environmental Dynamics in Arid Areas based on Markov Process—A Case Study of Wusu Oasis, Xinjiang *Xinjiang Agricultural Sciences* **48** 954-61 (In Chinese with English Abstract)

[6] Xiao D N, Chen W B and Guo F Q 2002 On the basic concepts and contents of ecological security *Chinese Journal of Applied Ecology* **13** 354-58 (In Chinese with English Abstract)

[7] Lu Y, Su W J and Hua C 2011 Analysis of Ecological Vulnerability in Zuojiang River Basin Based on Landscape Pattern and Ecosystem Sensitivity *Research of soil and water conservation* 2011 **18** 78-87 (In Chinese with English Abstract)

[8] Chen P, Fu S F, Wen C X, Wu H Y and Song Z X 2014 Assessment of Impact on Coastal Wetland of Xiamen Bay and Response of Landscape Pattern from Human Disturbance from 1989 to 2010 *Journal of Applied Oceanography* **33** 167-74 (In Chinese with English Abstract)

[9] Walz U, Stein C 2014 Indicators of Hemeroby for the Monitoring of Landscapes in Germany J. Nat. Conserv. **22** 279-89

[10] Chen A L, Zhu B Q, Chen D L, Wu Y H and Sun R H 2010 Dynamic Changes of Landscape Pattern and Eco-disturbance Degree in Shuangtai Estuary Wetland of Liaoning Province *China Journal of Applied Ecology* **21** 1120-28 (In Chinese with English Abstract)

# Extracting Vegetation Coverage in Dry-hot Valley Regions Based on Alternating Angle Minimum Algorithm

**M Y Yang[1], J Wang[1*] and Q Zhang[2]**
[1]China West Normal University, College of Land and Resource, Nanchong 637300, China;
[2]Key Laboratory of Digital Globe, Institute of Remote Sensing and Digital Earth of CAS, Beijing 100094,China

E-mail: wangjie308@mails.ucas.ac.cn

**Abstract.** Vegetation coverage is one of the most important indicators for ecological environment change, and is also an effective index for the assessment of land degradation and desertification. The dry-hot valley regions have sparse surface vegetation, and the spectral information about the vegetation in such regions usually has a weak representation in remote sensing, so there are considerable limitations for applying the commonly-used vegetation index method to calculate the vegetation coverage in the dry-hot valley regions. Therefore, in this paper, Alternating Angle Minimum (AAM) algorithm of deterministic model is adopted for selective endmember for pixel unmixing of MODIS image in order to extract the vegetation coverage, and accuracy test is carried out by the use of the Landsat TM image over the same period. As shown by the results, in the dry-hot valley regions with sparse vegetation, AAM model has a high unmixing accuracy, and the extracted vegetation coverage is close to the actual situation, so it is promising to apply the AAM model to the extraction of vegetation coverage in the dry-hot valley regions.

## 1. Introduction

Vegetation coverage refers to the percentage of vertical projection area of vegetation (including leaves, stems and branches) in the total area of the statistical area, which is not only a comprehensive quantitative indicator for the conditions of the ground surface covered by plant communities, but also a key factor to control soil erosion. At present, obtaining the vegetation coverage by remote sensing technology is the main method for obtaining the vegetation coverage information in a wide area,

* Corresponding author.

IOP Conf. Series: Earth and Environmental Science **74** (2017) 012019 doi:10.1088/1755-1315/74/1/012019

which is characterized by low cost, timeliness and accuracy, etc. Currently, there are many methods for the extraction of vegetation coverage through remote sensing images, mainly including empirical model method, vegetation index method, pixel decomposition model method, decision tree classification algorithm, and artificial neural network algorithm. Researchers like Li et al. (2011) have systematically studied the retrieval method for vegetation coverage based on hyperspectral data by estimating the vegetation coverage in the arid areas (vegetation coverage < 40%) through different NDVI and PLS(Partial Least-squares Regression)regression models, and with spaceborne hyperspectral image Hyperion as the data source [1]; researchers like Sun et al. (2006) have estimated the vegetation coverage in the Beiyunhe Valley by making use of the dimidiate pixel model, with the minimum NDVI value as soilNDVI and the maximum NDVI value as vegNDVI, and have achieved significant results [2]. By using MODIS and other remote sensing data, researchers like Hansen et al. (2002) have established a decision tree based on NDVI vegetation index, emissivity of red spectral band, etc., modified the decision tree that had been established by use of other data, and estimated the arbor coverage in central Africa with the modified decision tree and worked out a high accuracy of estimation [3]. The above studies provide a number of references for extracting the vegetation coverage through remote sensing images. However, in view of the feature that surface vegetation is sparse in the dry-hot valley regions, there is still lack of a simple and easy method for extracting the vegetation coverage in dry-hot valley regions. Therefore, in this paper, we will test the effect of and feasibility for extracting the vegetation coverage by means of Alternating Angle Minimum algorithm with MODIS image that can be easily obtained as the data source and with the typical dry-hot valley area in Yuanmou County as the research area.

## 2. Research area and remote-sensing image preprocessing

Yuanmou County is chosen as the research area. Located on the riverside of the Jinsha River, the county is under the jurisdiction of Yi Autonomous Prefecture of Chuxiong, Yunnan Province; situated in the northern part of middle Yunnan plateau with coordinates of 101°35′-102°06′E and 25°23′-26°06′N, the county has a dry and hot monsoon climate of south subtropics, with an annual average temperature of 21.9 °C, extreme maximum temperature of 42 °C, minimum temperature of minus 0.1 °C, and an annual evaporation 6.4 times of the annual precipitation, outlining a typical dry-hot valley region. The data used in this paper include the synthetized MODIS image over eight days during January 2004, February 2005 and March 2010 of Yuanmou County and Landast5 TM image over the same period, which was obtained through the geospatial data cloud. The data imaging time was in January 2004, February 2005 and March 2010, respectively. Because of the spring dry season in Yuanmou, less cloud cover, the image of the imaging effect is good, so choose the image of the season as the experimental material. The number and content of cloud for MODIS and TM image are shown in table 1.

IOP Conf. Series: Earth and Environmental Science **74** (2017) 012019 doi:10.1088/1755-1315/74/1/012019

**Table 1.** MODIS and TM images SN and contents of cloud.

| Image Type | MODIS Image | | TM Image | |
|---|---|---|---|---|
| | SN | Content of cloud | SN | Content of cloud |
| 2004 | MYD09A1.A2004017.h27v06.005 | No cloud | LT51300422004013BJC00 | No cloud |
| 2005 | MYD09A1.A2005033.h27v06.005 | No cloud | LT51300422005031BKT02 | No cloud |
| 2010 | MYD09A1.A2010073.h27v06.005 | No cloud | LT51300422010077BKT00 | No cloud |

Procedures for MODIS image processing: Firstly, the images in HDF format are converted to the standard ENVI format; the MYD09GA product is re-projected to be consistent with TM image coordinates. Secondly, wavebands are combined and arranged in ascending order based on the wavelength. Finally, the corresponding MODIS images to TM images are formed after shearing, as shown in figures 1a, 1b and 1c (all images are displayed in the red, blue and green band).

Procedures for TM image processing: Firstly, radiometric calibration to TM data is carried out to generate a radiation value, and the Gain and Bias coefficients are shown in the literature [4]. As the reflection of ground objects is easily affected by such factors as atmosphere and light, it is necessary to perform atmospheric correction to TM images. The FLAASH atmospheric model is used. The sub region of the research area are extracted, as shown in figures 2a, 2b and 2c (all images are displayed in the red, blue and green band).

As MODIS Terra data involves a large amount of cloud, tests are performed based on the 8-day synthesized data of MODIS Aqua. From table 1, it can be seen that MODIS images do not exactly match with the time of TM images, with the maximum time difference of 4 days. However, this does not affect the retrieval of vegetation coverage, because the temporal variation in vegetation coverage is not obvious during this period. In the absence of measured vegetation coverage data in Yuanmou County, TM data characterized by high spatial resolutions, relatively small mixed pixels and relatively high precisions of vegetation coverage extraction are adopted for verifying the vegetation coverage extracted from MODIS images.

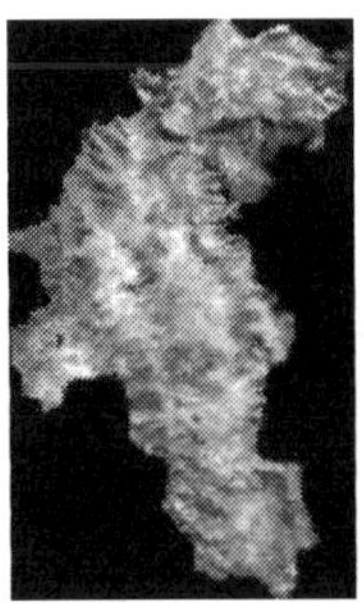

**Figure1a.** MODIS image in 2004.

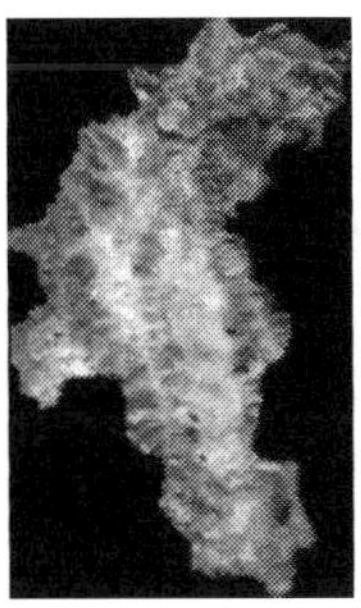

**Figure1b.** MODIS image in 2005.

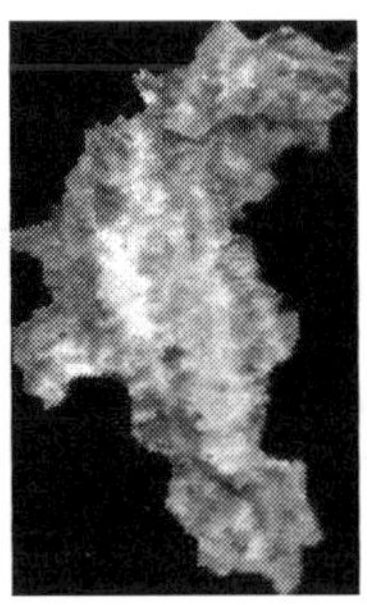

**Figure1c.** MODIS image in 2010.

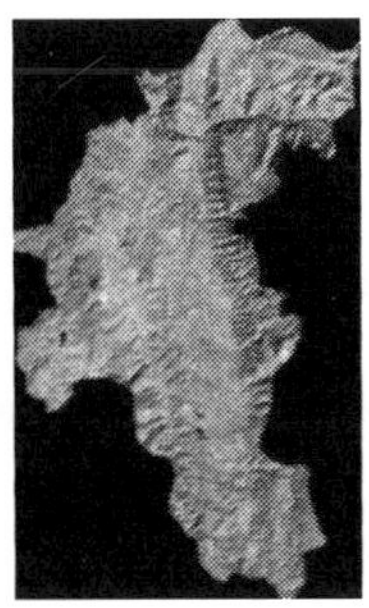

**Figure 2a.**TM image in 2004.

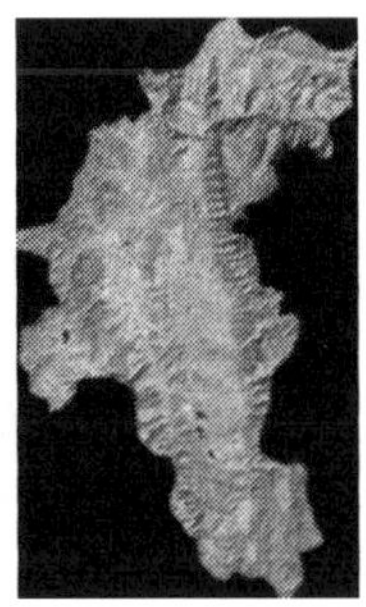

**Figure 2b.**TM image in 2005.

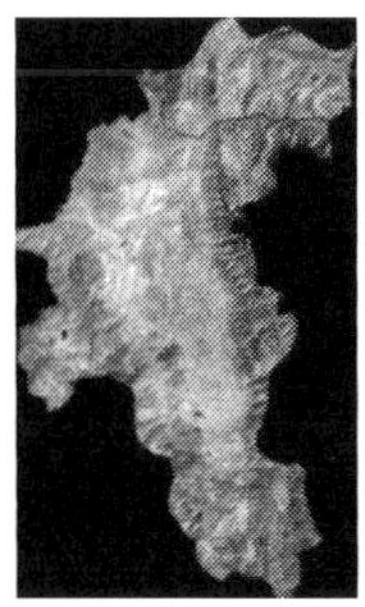

**Figure 2c.** TM image in 2010.

### 3. Alternating angle minimum method (AAM) endmember variability unmixing

Linear mixed model (LMM) is a spectral analysis model commonly used in remote sensing image processing. It is mainly used for unmixing the whole remote sensing image after selecting the spectrum of characteristic ground object as the endmember. Fully constrained least squares method (FCLS) [6] is a widely used hybrid pixel decomposition algorithm, as shown in equation (1):

$$x=\sum_{i=1}^{R}a_i m_i+\eta,\ \forall i:a_i\geq 0\ and\ \sum_{i}^{R}a_i=1 \tag{1}$$

where: $a$ denotes the coefficient, $R$ denotes the number of end elements, $m$ denotes the end element, and $x$ denotes the pixel.

The fixed endmembers are easily affected by the changes in physical characteristics of ground objects, atmospheric and lighting conditions during unmixing, thus resulting in spectral changes of the endmembers. In this case, there will be certain limitations of the FCLS algorithm, so the multiple-endmember spectral mixture analysis algorithm (MESMA) [7] is proposed to deal with the changes of endmembers. Although the changes of endmembers can be properly handled by MESMA algorithm, the computation efficiency of this algorithm is too low to ensure a high applicability. Due to such limitation in the MESMA algorithm, Alternating Angle Minimum algorithm [8] is adopted for improving the computation efficiency of the program.

If we interpret each spectrum as a d-dimensional vector spectral space [0, 1] and assume that spectral noise is distributed in a consistent manner, applying the maximum likelihood method to the linear mixture problem will result in least-squares problem, which is equivalent to finding out the minimum Euclidian distance target data point $X$ and mapping it onto a given subset $S$ of the spectral space:

$$\mathrm{y}=\mathrm{P_S}(\mathrm{x})\Leftrightarrow\forall \mathrm{z}|\in \mathrm{S}:\mathrm{z}-\mathrm{x}\geq \mathrm{y}-\mathrm{x} \tag{2}$$

If we only consider the constraint that the sum is 1, the unmixing problem is equivalent to making an orthogonal projection of target $x$ to the endmember on hyperplane where $H(E)$ passes through $E$. Reconstruction error is given from the orthogonal distance from $x$ to $H(E)$. If $F=E/\{e_p\}=(e_1,\mathrm{K},e_{p-1})$ and the standard geometric relationship is adopted, the orthogonal distance can be expressed as:

$$\|x-P_{H(E)}(x)\|=\|x-P_{H(F)}(x)\|\sin(\theta) \tag{3}$$

where,

$$\theta=\min(\alpha,\pi-\alpha) \tag{4}$$

$$\alpha=\sin^{-1}\left(\frac{\|x-P_{H(G)}(x)\|}{\|x-P_{H(F)}(x)\|}\right) \tag{5}$$

## 4. Results and comparison

In this paper, the processed TM images with high spatial resolutions are taken as the test reference,

IOP Conf. Series: Earth and Environmental Science **74** (2017) 012019 doi:10.1088/1755-1315/74/1/012019

and the root-mean-square error (RMSE) and correlation coefficient (R) index are adopted for evaluating the accuracy of extraction results of AAM. The original TM images are classified by the use of the support vector machine (SVM) algorithm, and the research area is classified into vegetation type and non-vegetation type according to the soil map and land use map of Yuanmou county, including six vegetation types. The SVM algorithm is embedded in ENVI software, radial basis function is used as the kernel function, with the value of the kernel function set as 0.167, the penalty parameter set as 100, the pyramid level set as 1, and the pyramid reclassification threshold set as 0.9. Visual correction is made to the classified images, and the 6 corrected vegetation types are integrated into one category and saved as the vegetation coverage of this region. As the spatial resolution of TM images is not consistent with that of MODIS images, therefore, it is necessary to carry out pixel clustering to the combined vegetation image, i.e. clustering its spatial resolution from 30M into 463.6M and making it consistent with the spatial resolution of MODIS images. By this way, the retrieval of vegetation coverage from TM algorithm in Yuanmou is determined (Figure 3. Retrieval of Vegetation cover fraction from the TM algorithm in: a. 2004; b. 2005; and c. 2010.). It is necessary to establish an endmember spectrum library in advance when AAM model is used for obtaining the vegetation coverage. The endmembers on MODIS images are generally divided into five categories: soil, rock, vegetation, water body and shadow area. Here, the endmembers are selected in an evenly-distributed manner on MODIS images based on the soil map, land use map and true value of vegetation after pixel clustering, with 10 endmembers selected for each category. AAM model is used for extracting the vegetation coverage of MODIS images for three time periods, and the results are shown in figures 4a, 4b and 4c（Figure 4. Retrieval of Vegetation cover fraction using the AAM in: a. 2004; b. 2005; and c. 2010.）. According to the comparison between figure 3 and figure 4 and based on the accuracy evaluation results shown in table 2, it can be seen that AAM has a high correlation coefficient and a small root mean square error, and the extraction results are close to the actual vegetation coverage, so the situation of local vegetation coverage can be retrieved correctly.

**Table 2.** Comparison of extraction accuracy indexes of AAM model.

| Assessment Index | RMSE | R |
|---|---|---|
| 2004 | 0.1585 | 0.8350 |
| 2005 | 0.1447 | 0.8171 |
| 2010 | 0.1337 | 0.8472 |

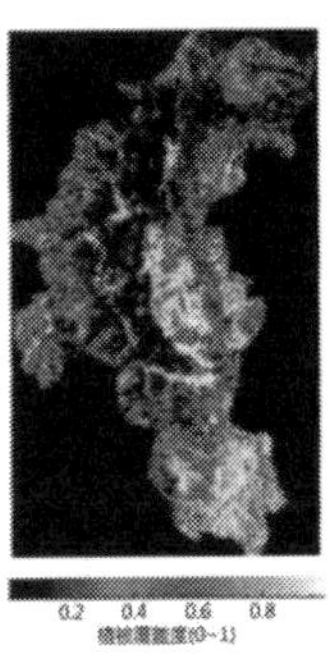
**Figure 3a.**

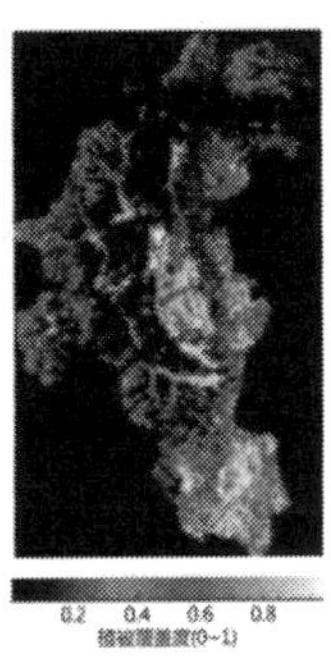
**Figure 3b.**

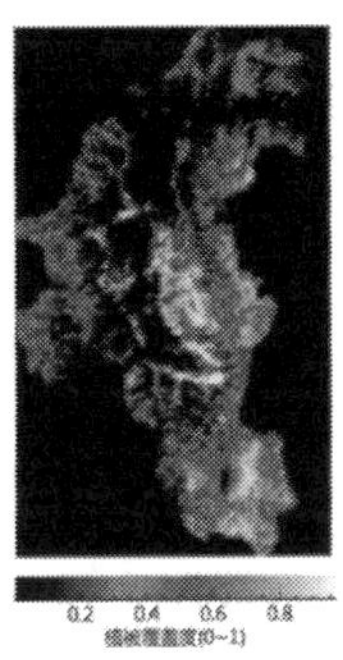
**Figure 3c.**

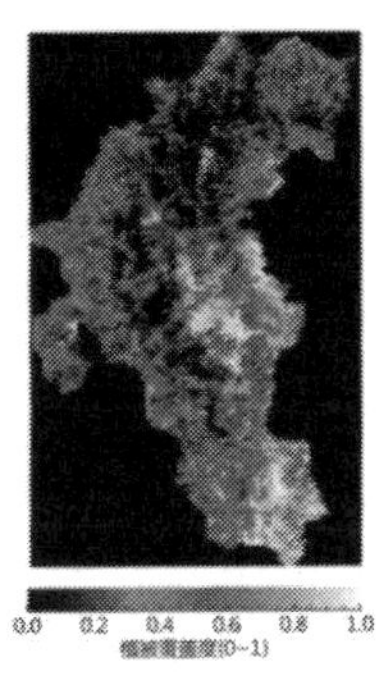
**Figure 4a.**

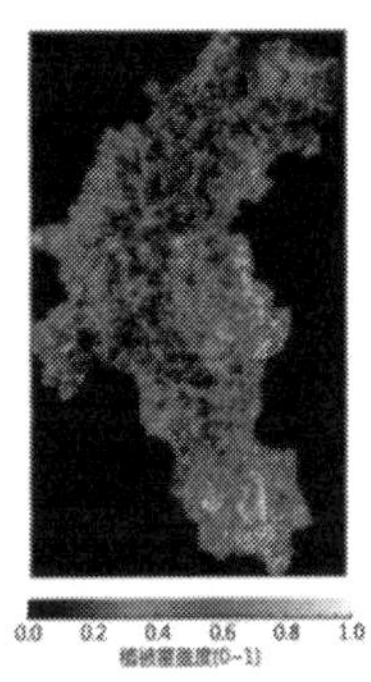
**Figure 4b.**

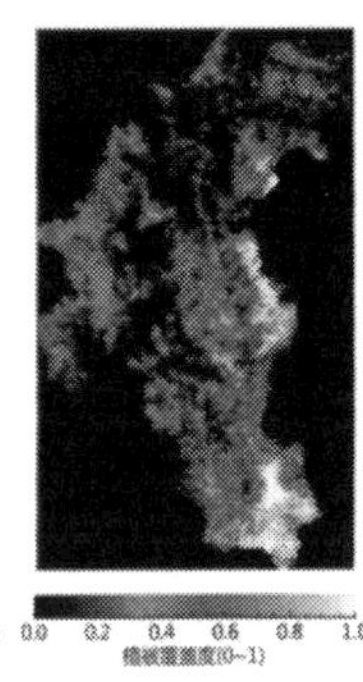
**Figure 4c.**

2004. 2005. 2010. 2004. 2005. 2010.

## 5. Conclusion

In this paper, AAM model is used for extracting the vegetation coverage based on MODIS images, and accuracy comparison is made for the results. The results show the feasibility for applying the unmixing method to the MODIS remote sensing images and its application prospects, and the following conclusions can be drawn:

(1) In view of the extraction of vegetation coverage, the use of AAM model can effectively avoid the difficulty in obtaining pure pixels from MODIS images with dimidiate pixel model and three-band gradient difference method, reducing the error in the extraction process, and thus more accurate extraction results can be obtained.

(2) AAM model can effectively solve the problems that the vegetation coverage in the dry-hot valley regions is low and the contribution of exposed soils to pixel spectrum may easily cover up the contribution of vegetation to the pixel spectrum. The algorithm is more applicable to the sparse vegetation areas than the pixel decomposition model, so it is an effective method for extracting the vegetation coverage in the dry-hot valley regions. With MODIS images as the data source, which can be obtained easy, the method can be widely applied.

The AAM algorithm has a good effect on the extraction of vegetation coverage in arid area, but there is still great room for improvement. In the future, we will try to extract the endmember automatically instead of manually selecting the pure pixels, and increase the spatial neighborhood information of the AAM algorithm to solve the problem of low smoothness and continuity.

## References

[1] Li X S, Li Z Y and Gao Z H 2011 Estimation of Vegetation Cover in Desertified Regions from Hyperion Imageries Using NDVI and Partial Least Squares Regression *Journal of Desert Research* **31** 162-67 (In Chinese with English Abstract)

[2] Sun J H, Liu X M and Li Y G 2006 Estimation of Vegetation Fraction in Beiyunhe District by Remote Sensing *Research of Soil Water Conservation* **13** 97-99 (In Chinese with English Abstract)

[3] Hansen M C, De Fries R S and Townshend J R G 2002 Towards an Operational MODIS Continuous Field of Percent Tree Cover Algorithm: Examples Using AVHRR and MODIS Data *Remote Sens. Environ.* **83** 303-19

[4] Chander G, Markham B L and Helder D L 2009 Summary of Current Radiometric Calibration Coefficients for Landsat MSS, TM, ETM+, and EO-1 ALI Sensors *Remote Sens. Environ.* **113** 893-903

[5] Chen Q and Chen Y F 2005 A Study on Estimation of Vegetation Fractionby Using QuickBird Imagery *Forest Research* **4** 375-80 (In Chinese with English Abstract)

[6] Heinz D C and Chang C 2001 Fully Constrained Least Squares Linear Spectral Mixture Analysis Method for Material Quantification in Hyperspectral Imagery *IEEE Trans. Geosci. Electron.* **39** 529-45

[7] Roberts D A, Gardner M and Church R 1998 Mapping Chaparral in the Santa Monica

Mountains Using Multiple Endmember Spectral Mixture Models *Remote Sens. Environ.* **65** 267-79

[8] Hu M G and Wang J F 2010 Mixed-pixel Decomposition and Super-resolution Reconstruction of RS image *Progress In Geography* **29** 747-56 (In Chinese with English Abstract)

[9] Keshava N and Mustard J F 2002 Spectral Unmixing *IEEE Signal Proc. Mag.* **9** 44-57

[10] Bioucas Dias J M, Plaza A and Dobigeon N 2012 Hyperspectral Unmixing Overview: Geometrical, Statistical, and Sparse Regression-based Approaches *IEEE J. STARS* **5** 354-79

[11] Zare A and Ho K 2014 Endmember variability in hyperspectral analysis:Addressing spectral variability during spectral unmixing *IEEE Signal Proc. Mag.* **31** 95-104

IOP Conf. Series: Earth and Environmental Science **74** (2017) 012020 doi:10.1088/1755-1315/74/1/012020

# Temporal and spatial characteristics of water resources in the Yeerqiang River Basin based on remote sensing

**Q Y Ran[1,2], L Y Bai[2], J Z Feng[3*], Y M Yang[1], M Q Guo[2], H L Li[2], Q Zhang[2], P Zhang[2] and D Cao[2]**

[1]College of Hohai, Chongqing Jiaotong University, Chongqing 400074, China

[2]Institute of Remote Sensing and Digital Earth, Chinese Academy of Sciences, Beijing 100094, China

[3]Key Lab of Agri-information Service Technology, Ministry of Agriculture of P. R. China, Agricultural Information Institute, Chinese Academy of Agricultural Sciences, Beijing 100081, China

E-mail: baily@radi.ac.cn

**Abstract.** Surface water resources play an important role in the economic and social developments as well as the protection of natural ecological environment in the Yeerqiang River Basin. Based upon the six stages of land use data from 1990 to 2015, the temporal and spatial variation of surface water resources in the Yerqiang River Basin have been explored and analyzed. The results show that: (1) From 1990 to 2015, the area of natural landscape initially increased and then decreased, while the area of artificial landscape increased, which caused a slight increase in the land use degree in the study area. (2) The dynamic changes of water and glacier areas are somewhat consistent over the past 25 years, with a sharp decline between 2005-2010 and a small increase in the remaining years. The dynamic changes in areas of non-glacial water were moderate, with decrease in area of 9 $km^2$ from 1990 to 2015. The beach area decreased, and the other water sub-classes initially increased and then decreased. (3) Over the past 25 years, the proportion of unchanged water area is 73.22%, the transfer-out proportion is 19.19%, and the transfer-in proportion is 7.59%. Generally, water types transferred to grassland and unused land. Additionally, significant transfers were observed for the conversions between glaciers and woodland, conversions between canal, lake, reservoir and beach, and conversions between beach and farmland.

## 1. Introduction

The Tarim River Basin, an area of Xinjiang with a large population, plays a significant role in the regional economic activity, ecological function, and environment protection [1]. However, ecological and environmental problems, such as water resources shortages, desert riparian forest degradation, and land desertification have presented an increasingly serious situation in the region [2-3]. Surface water resources are significant for the stabile and healthy development of the ecological environment in the basin. The Yerqiang River is the longest headstream of the Tarim River, accounting for 3.6% of the all Tarim River runoff recharge; therefore, the surface water resources in the Yerqiang River Basin are representative of the surface water resources in the Tarim River Basin. Thus, we selected the Yeerqiang River Basin as the study area for this research and conducted dynamic monitoring of land use and surface water resources from 1990 to 2015. In this way, we hope that our study can assist in environmental conservation and government regulation in this region.

## 2. Data and methodology

### *2.1. Study area*

The Yerqiang River is located in the western region of the Tarim Basin and is one of the main origins of the Tarim River, with river length of 1179 km. The Yerqiang River Basin is bordered on the west by the Tuohulake Desert, on the east by the Taklimakan Desert, and surrounded to the north and south by the Tianshan and Kunlun mountains (figure 1). The basin extends across four regions, including Kashi, Hotan, Kizilsu Kirgiz Autonomous Prefecture and Akesu. The total area is 9.89×104 km$^2$, of which the mountains account for 61.5% and the plains account for 38.5% [4-5]. The Yerqiang River Basin is located in an arid region of Northwest China and belongs to the warm temperate continental plateau climate, with low precipitation and strong moisture evaporation [6].

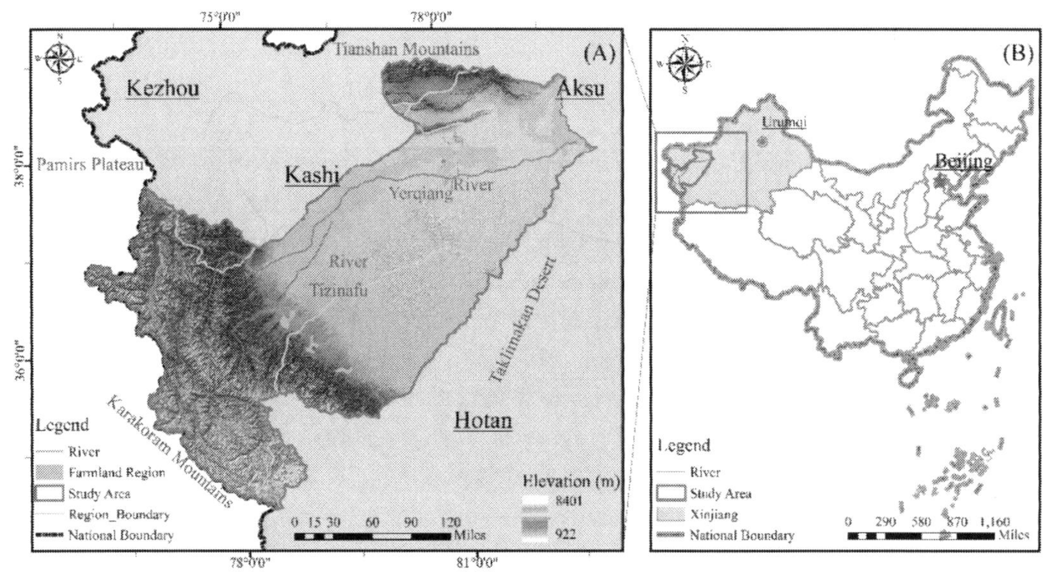


**Figure 1.** Location (B) and scope (A) of the study area.

### *2.2. Data*

The data used in this study are as follows: (1) China's rivers, administrative divisions and other basic geographic vector data; (2) Land use data (30 m raster data) of Xinjiang from 1990 to 2015. The data were generated by manual visual interpretation, using the Landsat TM/ETM/OLI remote sensing images as the data source. The data include six primary land use types (farmland, woodland, grassland, water, residential land and unused land) and twenty-five subclass land use types. The sub-class land use types of water include canal, lake, reservoir and pond (referred to as reservoir), permanent ice and snow (referred to as glacier), and beach.

### *2.3.Methodology*

In order to research land use, especially the temporal and spatial variation of surface water resources, the land use degree index calculation formula was used to calculate the land use degree index, as presented by Liu et al. [7-8]. This index can be used to quantitatively reveal the comprehensive land use level and has an important role in reflecting the change of land use degree. Additionally, in order to determine the intensity of surface water resources changes, this study used a single land use dynamic degree [9]. Finally, a transfer matrix was used to analyze the transfer features between surface water resources and other land use types in order to elucidate the transfer source and destination of surface water resources [10].

## 3. Result and discussion

### *3.1. Regional land use change*

The grassland and unused land have a large area proportion in the study area. These two land types account for 82.01% of all land use areas. Water areas account for 9.78%, followed by farmland, woodland and residential land accounting for the least, with areas 6.89%, 0.98% and 0.34%, respectively (figure 2). The types that most closely related to human activities are farmland and residential land. The area changes of these two types showed a significant increasing trend. Among them, the change in area of farmland shows a smaller reduction from 1990 to 1995, but it increased over the next two decades, with a total increase of 4295 $km^2$ (figure 2). With respect to farmland, the area of residential land showed a slow growth trend, with a total increase of 161 $km^2$ from 1990 to 2015 (figure 2). The combination of these two factors causes an increase in human consumption of regional water resources and, therefore, changes the temporal and spatial distribution pattern of regional surface water. The woodland, grassland and unused land in the study area were affected less by human activities but affected more by natural environment. Woodland was the low proportion type in the study area. The woodland area first increased and then decreased, with a total increase value of 89 $km^2$ over 25 years (figure 2). The change in grassland area increased from 1990 to 1995, but later it decreased a total of 5941 $km^2$ over 20 years (figure 2). The area of unused land fluctuated irregularly from 1990 to 2015 (figure 2).

The land use degree index can quantitatively describe the comprehensive level of regional land use and indirectly reflect the degree of regional water consumption. The land use degree index of the study area was affected by the combined effects of land use/cover changes and initially showed an increasing trend, followed by a decrease. The total increase in index value was 1.77 over the last 25 years; the minimum value was 151.15 in 1990 and the maximum was 154.48 in 2005 (figure 2). In addition, although the land use degree in the study area increased, the average value of the land use degree index was 152.63, which is significantly smaller than the limit value (400). On one hand, it is closely related to the high proportion of unused land (accounting for 54.64% of the total) in the study area, and on the other hand, it shows that the land use degree and development intensity of this study area remained relatively low between natural and agricultural land use.

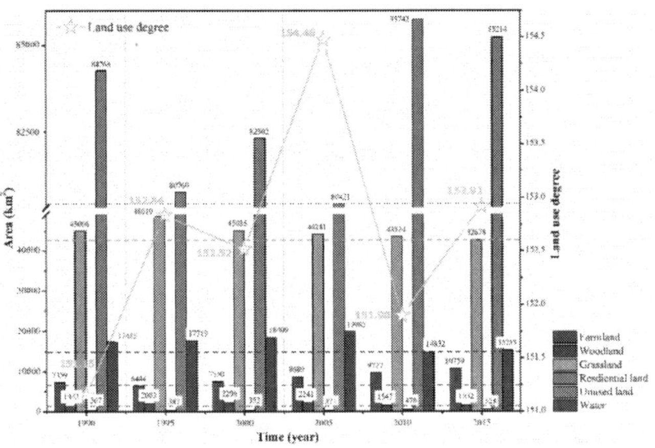


**Figure 2.** Changes in regional land use in the study area from 1990 to 2015.

### *3.2. Dynamic change characteristics of water*

The land use single dynamic degree can be used to study the dynamic changes of the water areas. More specifically, it can reflect the severe degree (amplitude and velocity) of the change in water areas in the study area. The glacier category is the dominant water type in the study area, and it accounts for 92.28% of the total water area. Therefore, the overall change in water area was consistent with that of the glacier area. The glacier category showed a slight increase in area from 1990 to 2005, and later it sharply declined, with an area reduction of 5097 km$^2$ from 2005 to 2010. Although, in the last five years, the area slightly increased (figure 3). Compared with the glacier, the change non-glacial water area was relatively slow, and it fluctuated, with a total area decrease of 9 km$^2$ over the last 25 years (figure 3). The canal area also fluctuated, reaching a maximum value of 413 km$^2$ in 1995 and a minimum value of 243km$^2$ in 2005. The total area increased to 25 km$^2$ for the canal area (figure 3). The lake area first increased, and then decreased from 1990 to 2010. Later, the lake area increased sharply to 166 km$^2$ from 2010 to 2015, and the dynamic degree reached 30.74% (figure 3). The reservoir area significantly increased after an initial decreasing trend, and the minimum value was 206 km$^2$ in 1990. The maximum value for the reservoir area was 360 km$^2$ in 2000 and 253 km$^2$ in 2015 (figure 3). The beach area increased slightly from 1995 to 2000, and yet in the rest of the periods the areas were reduced. The greatest reduction was observed from 1990-1995, with an area decrease of 200 km$^2$. Ultimately, the beach area showed a total reduction of 240 km$^2$ over the last 25 years (figure 3).

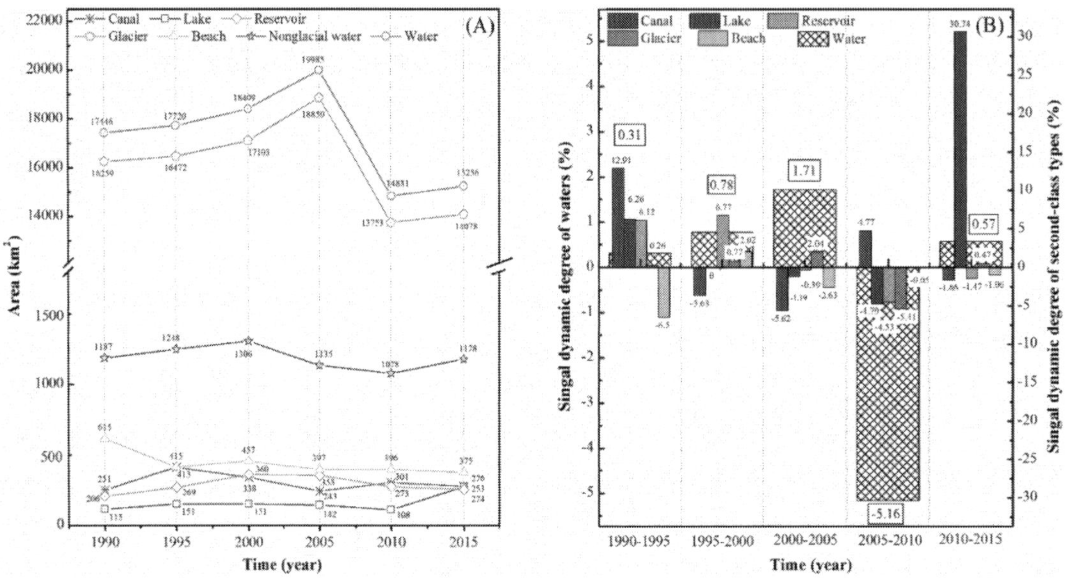

**Figure 3.** The dynamic change of water in study area from 1990 to 2015.

### *3.3 .The transfer characteristics of water areas*

Through quantitative analysis of the dynamic change in water area, we know that the water area first increased and then decreased from 1990 to 2015. The water area decreased to 2190 km$^2$; the glacier area decreased to 21810 km$^2$; and the non-glacial water area decreased to 90 km$^2$. In order to identify the location and the associated land use/cover types of these changes, the transfer rule between water types and other types in the study area were obtained by statistical analysis based on the six stages of land use data. In the past 25 years, the proportion of non-change water area is 73.22%, and total transfer water area is 26.78%, of which 19.19% is water transferred into other types and 7.59% are other types transferred into water (figure 4). The areas in water categories transferred mainly into grassland and unused land, and the transfer proportions were 30.14% and 66.74%, respectively. The total was 96.88%, and the remaining water categories were mainly transferred into farmland, with a proportion of 2.86% (figure 4). Similarly, the types that transferred into water were mostly grassland and unused land, and the transfer proportions were 26.76% and 71.03%, respectively. The total was 97.79%, and the remaining water categories were mainly transferred from farmland with a proportion of 1.66% (figure 4).We observed the transfer changes between water and other types mainly occurred in the glacier region, and those types were mainly grassland and unused land. In addition, in the low altitude plain area, a large number of non-glacial water converted with farmland and woodland, with the exception of grassland and unused land. The transfer law between water and other types in local research stages is more consistent with that of 25 years. The proportion of the un-changed water area decreased from 85.26% in the1990-1995 stage to 72.05% in the 2010-2015 stage. The proportion of the transfer area of water in each local period is large, but there is no uniform law. However, there is a phenomenon that the proportion of the water transferred into farmland is greater than the farmland transferred into water, and this indicates that water areas were developed into farmland in the past 25 years.

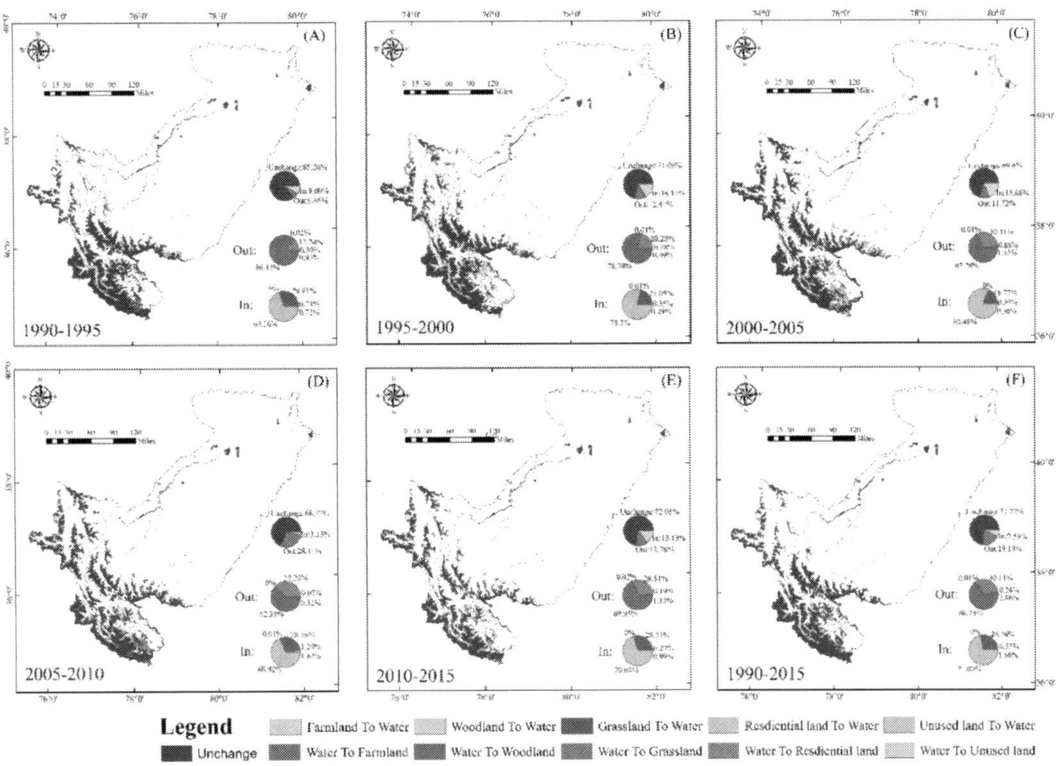


**Figure 4.** The transfer destination of water from 1990 to 2015 in the study area.

*3.4 .The transfer characteristics of water sub-classes*

Before, we calculated the transfer rule between water and other types, but the transfer relationships between sub-class water and other types are not clear. Therefore, in order to further analyze the transfer characteristics (transfer destination and source) of sub-class water, we constructed an area transfer matrix. In all the sub-classes of water, the glacier category is considered solid surface water, and its area was larger than other water sub-classes. It also plays an important role in the regional runoff and reservoir water supply. In the past 25 years, the glacier area was unchanged at 12912 $km^2$, and the transfer area of glacier to other types is 3345 $km^2$. Of the 3345 $km^2$, 961 $km^2$ and 2384 $km^2$ were transferred into grassland and unused land, respectively (figure 5). In addition, the glacier category is exclusively transferred from grassland and unused land with 206 $km^2$ and 959 $km^2$, respectively (figure 5). In the five local stages, the types that mutually transferred with glacier are still dominated by grassland and unused land; however, some transfer with canal and woodland categories was observed. Canal refers to man-made or natural forms of water reservoirs, with linear characteristics and permanent location below the water surface curve. As the main type of sub-class non-glacial water, canals are important representatives of regional water resources. In the past 25 years, the area of the canal category was unchanged at 105 $km^2$, and the net conversion area from woodland, grassland and unused land to canal are 2 $km^2$, 29 $km^2$ and 16 $km^2$, respectively. The net conversion area from canal to farmland, lake and beach are 3$km^2$, 3$km^2$ and 17$km^2$, respectively (figure 5). The transfer relationships between canal and other types in local research stages are more consistent with that of the past 25 years. The main types mutual transferred with canal were grassland and beach. Lakes and reservoirs are both water storage facilities, and the difference between them is the origin of formation: natural or anthropogenic. Lake is irregular in boundary shape. Reservoirs are a kind of surface water body, which are constructed with obvious artificial features, hydraulic structures and dams. In the past 25 years, the area of lakes remained constant at 94 $km^2$, and the transfer source was mainly grassland, with a net transfer-in area of 23 $km^2$. The transfer destination was mainly beach, with a net transfer-out area of 111 $km^2$. The area transferred to other types was small and no more than 10 $km^2$ (figure 5). Compared with the transfer changes of the lake, the transfer changes of the reservoir were more severe. In the past 25 years, reservoir area remained constant at 117 $km^2$, and the transfer destination was mainly from reservoir to farmland and lake, with net transfer-out areas of 31 $km^2$and 18 $km^2$, respectively. The transfer sources were mainly grassland and beach, with a net transfer–in area of 29 $km^2$and 63 $km^2$, respectively (figure 5). In the local stages, there were more transferred areas between reservoir and beach, and beach predominantly transferred to lake. As one of the sub-classes of water, beach is defined as the land between the water surface curve of flood season and dry season. For most of the year, it is a surface area without water; therefore, beach is special compared with the other subclass water types. In the past 25 years, the beach category remained constant with an area of 228 $km^2$, and beach areas transferred into lake, reservoir and farmland were 111$km^2$, 63 $km^2$ and 45 $km^2$, respectively (figure 5). There are also 35 $km^2$ of net beach transferred to grassland, and the transfer area to woodland and unused land were small (figure 5). It is interesting to note that only canal area transferred into beach, with an area of 17 $km^2$.

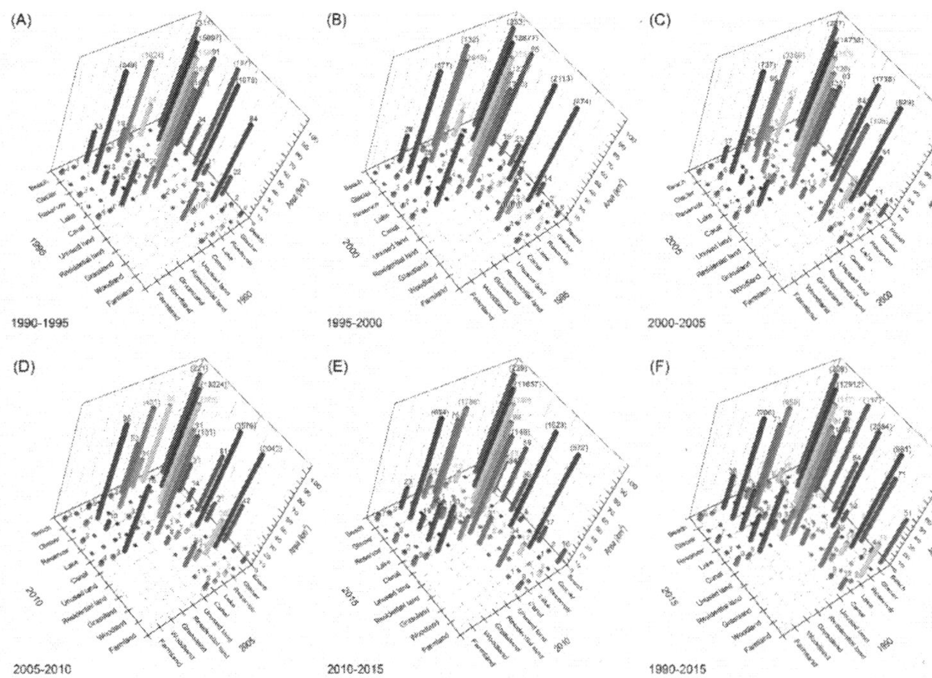


**Figure 5.** The transfer destination of sub-class water from 1990 to 2015 in the study area.

## 4. Conclusions

In this study, we quantitatively analyzed the dynamic changes of surface water resources in the Yeerqiang River Basin from 1990 to 2015. This study revealed surface water resources changed greatly over this time period, especially the transition changes in grassland, unused land and farmland. The changes were significantly influenced by human activities, in particular agriculture activities. The results of this study can offer useful insight to land use and surface water resources conservation as well as provide valuable information to policy makers, promote the implementation of the "The Belt and Road" program, and assist in regional environmental conservation and sustainable development.

**Acknowledge**

The authors would like to thank Western Light Talent Culture Project (YBXM-2014-05), the Natural Science Foundation of China (NSFC; Grant Nos. 41171340 and 41101390), and the Science and Technology Innovation Program (for the Agricultural Information Institute) of the Chinese Academy of Agricultural Sciences for supporting this work.

## Reference

[1] Wu L, Hou X Y and Xu X L 2013 Land Use and Landscape Pattern Changes in Coastal Areas of Shandong Province, China *Transactions of the Chinese Society of Agricultural Engineering* **29** 207-16 (In Chinese with English Abstract)

[2] Chen Z S, Chen Y N and Xu C C 2011 Change Trend and Prediction of Annual Runoff in Mainstream Area of the Tarim River in Recent 50 Years *Arid Land Geography* **34** 43-51 (In Chinese with English Abstract)

[3] Xu Y F 2013 Research of Water Resources Carrying Capacity in Tarim River Basin under the Climate Change (Zhengzhou: Zhengzhou University) p 84

[4] Li M 2006 Research on Exploitation of Reserved Cultivated Resources in Yeerqiang River based on GIS and RS (Urumchi: Xinjiang University) p 74

[5] Sun B G, Shen Y P and Wang G Y 2008 Changes in Runoff and River Sediments in the Upper Reaches of Yarkent River, Karakorum during 1954-2007 *Journal of Glaciology and Geocryology* **30** 1068-72 (In Chinese with English Abstract)

[6] Qin X W 2007 The Characteristic Water Level Optimal Choice and Benefits analysis of Aertashi Reservoir (Xi'an: Xi'an University of Technology) p 74

[7] Liu J Y 1992 Land Use in Tibet Autonomous Region (Beijing: Science Press)

[8] Gao Z Q, Liu J Y and Zhuang D F 1999 The Research of Chinese Land-Use/Land-Cover Present Situations *Journal of Remote Sensing* **3** 3134-38 (In Chinese with English Abstract)

[9] Zou B, Yang Y Q and Xue Y D 2011 Analysis on Dynamic Changes of Land Types and Stands in Pearl River Delta Based on RS and GIS *Forest Resources Management* **6** 110-16 (In Chinese with English Abstract)

[10] Meng Q and Liu X H 2014 Dynamic Change of Land Use in Fengdu County *Journal of Shanxi Agricultural Sciences* **42** 1225-29 (In Chinese with English Abstract)

IOP Conf. Series: Earth and Environmental Science **74** (2017) 012021 doi:10.1088/1755-1315/74/1/012021

# Temporal and Spatial Variations of Vegetation Cover in Xinjiang from 2002 to 2015 and Their Response to Climate

**S S Liu[1], Q Zhang [1*], X C Li [2], W J Song [1], J N Yang[1] and X J Liu[2]**
[1] Institute of Remote Sensing and Digital Earth Chinese Academy of Sciences, Beijing 100094, China;
[2] College of Information Science and Engineering, Shandong Agricultural University, Taian 271018, China

E-mail: zhangqing01@radi.ac.cn

**Abstract.** In this paper, the dataset of normalized difference vegetation indexes (NDVIs) in the arid region in Xinjiang from 2002 to 2015 as well as the climate data from 52 meteorological stations are utilized and the temporal and spatial variations of NDVI in recent years and their response to temperature and precipitation are analyzed in combination with various methods such as the maximum value analysis, correlation analysis and spatial analysis. It is concluded that in the past 14 years, the annual maximum NDVIs of Xinjiang presented a moderate rising tendency; Under the influence of the global background, the temperature and precipitation also showed different degrees of increase, which showed a significant increase in temperature. The annual maximum NDVI had a significant correlation with the annual precipitation (correlation coefficient: 0.634), but no obvious correlation was identified between the annual maximum NDVI and the annual average temperature (correlation coefficient: 0.279). To this end, regarding to the climatic factors, the influence of precipitation on the vegetation cover is higher than that of temperature.

## 1. Introduction

Located in the arid and semi-arid climatic zone of Eurasia, Xinjiang is an area with scarce vegetation, severe water and soil losses as well as extremely-fragile ecological environment. The climatic factors are principal elements leading to vegetation changes. In particular, temperature and precipitation have significant influence on the growth, distribution and carbon budget function of the vegetation [1]. Valuable explorations have been made by researchers on how to study land cover/vegetation dynamics with NDVI coupling with climatic factors. Gao and Liu(2016) [2] found that the vegetation index of the arid region of northwest China had little correlation with the precipitation and temperature, while Yan and Dai (2013) [3] pointed out that different types of vegetation had distinct responses to climatic factors, but the NDVIs of all vegetation did have significant positive correlation with the temperature and precipitation. Due to different temporal and spatial scales and analytical methods adopted for the study areas, previous studies drew distinct opinions and conclusions, and most of the study and analytical methods are simplex. In this paper, the temporal and spatial variations of NDVIs of the arid region in Xinjiang in the past 14 years as well as their response characteristics and sensibility to precipitation and temperature are analyzed, and the occurrence of extreme climatic events in Xinjiang in recent years as well as their influences on NDVI in allusion to the interannual variation of climatic factors are explored, so as to reveal the sensitivity of vegetation cover variations in the arid region in Xinjiang in the past 14 years to major climatic factors in the context of global warming and to provide scientific basis for guiding the oasis development and ecological environment construction in the arid region in Xinjiang with specific pertinence.

## 2. Data and study areas

---

* Corresponding author.

### *2.1. Overview of the study areas*

Xinjiang (73°20' ~ 96°25'E, 34°15' ~ 49°10'N) is located in the border area of northwest China with a total area of 1.66 × $10^6$ km2. It contains several major topographic units, such as the Altai Mountains, the mountainous region in the west of Dzungaria, Dzungarian Basin, Tianshan Mountains, Tarim Basin and Southern Mountain Area, which endows Xinjiang with a regional structure of "two basins embedded in three mountains". With scarce forests and low vegetation coverage (1.68%), the large area of desert, Gobi and open ground as well as the intensified land desertification year by year, Xinjiang has an extremely fragile ecological environment [4].

### *2.2. Data preparation*

*2.2.1. Acquisition of MODIS data.* The NDVI data used in this paper are collected from the data product on monthly vegetation index of MODIS MOD13A from 2002 to 2015 from the official website of NASA (resolution ratio: 1 KM × 1 KM), and the annual maximum NDVI is calculated by Maximum Value Composite(MVC).
*2.2.2. Meteorological data.* The data of the temperature and precipitation from 2002 to 2015 are collected from the annual average temperature and precipitation data from 2002 to 2013 and the monthly average temperature and precipitation data from 2014 to 2015 from 52 basic meteorological stations, which are provided by the Meteorological Information Center of Xinjiang Uygur Autonomous Region; the annual average temperature and precipitation data are obtained by processing the aforesaid data.

## 3. Study methods

### *3.1. Spatial interpolation method*

Inverse distance interpolation: the formula for inverse distance interpolation for the attribute z=z (x, y) of an arbitrary point (x, y) in the space is defined as follows:

$$\hat{z} = \sum_{i=0}^{n} \frac{1}{d_{\alpha}} z_i \qquad (1)$$

Where: $d_{\alpha}$ is the distance between the interpolated point and the first point in the field.
Kriging interpolation: the formula for Kriging interpolation for the attribute z=z (x, y) of an arbitrary point (x, y) in the space is defined as follows:

$$\hat{z}(s_0) = \sum_{i=1}^{n} \lambda_i z(s_i) \qquad (2)$$

Where: $z(s_i)$ is the measured value at position $i$ ; $\lambda_i$ is the unknown weight of the measured value at the $i$ position; $s_0$ is the mredicted position; $N$ is the measured value.

### *3.2. Correlation analysis method*

In this paper, the sensitivity of NDVI to corresponding climatic factors (including temperature and precipitation) is analyzed through correlation analysis. The formula for correlation analysis and calculation is provided as follows:

$$R_{xy,z} = \frac{R_{xy} - R_{xz}R_{yz}}{\sqrt{(1-R_{xz}^{2})(1-R_{yz}^{2})}} \qquad (3)$$

Where: n is the cumulative number of years in the monitoring period, X and Y are the 2 variables of the correlation analysis;and $\bar{x}$ , $\bar{y}$ are The average value of the sample values

## 4. Results and discussion

### *4.1. Analysis on interannual variation characteristics of vegetation cover*

The annual maximum NDVI represents the NDVI value in the period when the vegetation is the most exuberant. Changes in NDVI may reflect the interannual vegetation variation caused by climatic and human factors. In this paper, analysis is made on the interannual variation of the annual maximum of NDVI from 2002 to 2015 through Excel.

(1) The annual maximum NDVIs of Xinjiang from 2002 to 2015 fluctuated between 0.1638 and 0.1853, At 2008, the maximum of NDVI reach the minimum.

(2) Interannual variation tendency: phase of rapid growth (2009-2010; 2014-2015), phase of rapid decrease (2002-2004; 2007-2009; 2013-2014) and phase of moderate growth (2004-2006; 2010-2013).

(3) In the past 14 years, an overall rising tendency was found in the annual maximum NDVIs of the arid region in Xinjiang. As a symbol of vegetation cover variations, it indicates that the overall ecological environment of the arid region in Xinjiang has been inproved.

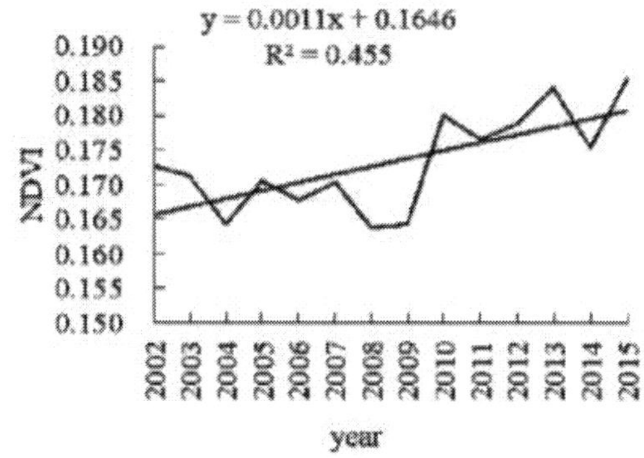


**Figure 1.** interannual Variation of Annual Maximum NDVI of Xinjiang from 2002 to 2015

### *4.2. Temporal variation of climatic factors and the influence of extreme climatic events on vegetation*

(1) The curve of interannual variation of annual precipitation in Xinjiang (figure 2) showed a slight upward tendency of the annual precipitation in the arid region in Xinjiang from 2002 to 2015.

(2) The minimum precipitation occurred in 2008, when the second severe drought disaster in the history of meteorological record happened. The disaster caused significant damages to the vegetation cover as well as human production and living activities in Xinjiang. It was also the reason for the obvious and rapid decrease tendency of annual maximum NDVI of Xinjiang to occur in 2008 and 2009.

(3) Embraced a rapid growth in 2009-2010, the precipitation in Xinjiang reached 185 mm in 2010. The curve (figure 3) shows an obvious increasing tendency (including a periodic decreasing tendency from 2007 to 2012) of the annual average temperature of Xinjiang. The lowest annual mean temperature (6.18°C) occurred in 2012. The annual mean temperature rose again in 2012-2015,and then the highest annual average temperature occurred in 2015.

(4) As for the growth rate, the annual precipitation increased 0.407 mm/year from 2002 to 2015 (figure 2), while the annual temperature increased 0.079°C/year during the same period (figure 3). This indicates that in the context of global warming, the tendency towards warm and wet climate has been relatively obvious since the 21st century in the arid region in Xinjiang.

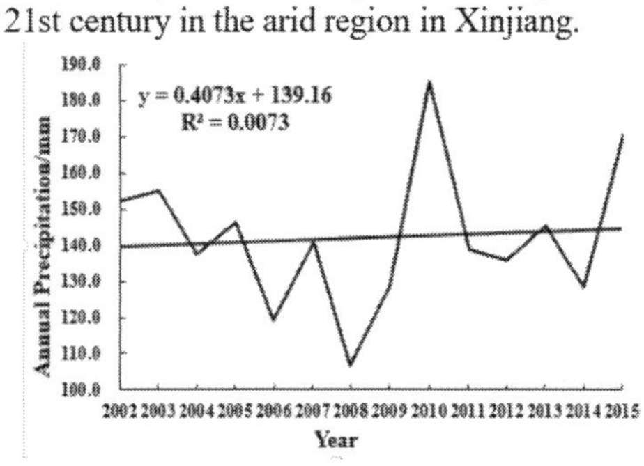


**Figure 2.** Interannual Variation of Annual

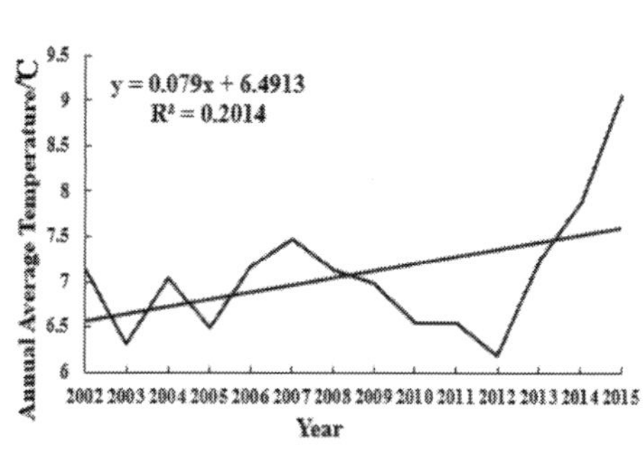


**Figure 3.** Interannual Variation of Annual

Precipitation from 2002 to 2015. Average Temperature from 2002 to 2015.

*4.3. Spatial pattern of temperature and precipitation*

(1) The precipitation showed a decreasing trend from northwest to southeast (figure 4), the middle of the Tarim Basin had a lowest annual precipitation of 50 mm/year.

(2) The spatial distribution of regional annual average temperature showed a decreasing trend character from south to north (figure 5), where the lowest annual average temperature was -3.5°C and the highest was 15.7°C.

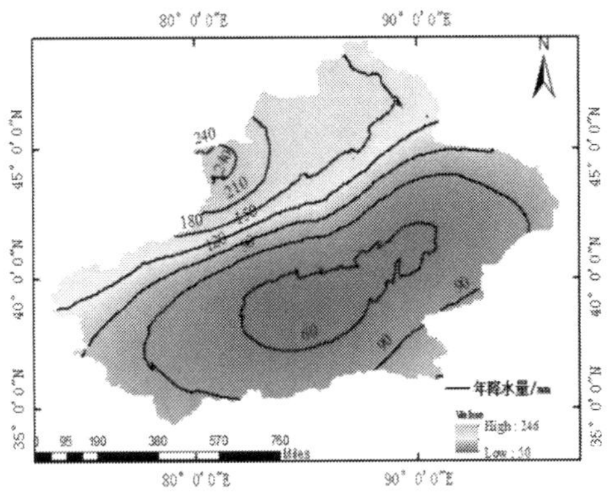


**Figure 4.** Spatial Distribution of Annual Average precipitation from 2002 to 2015

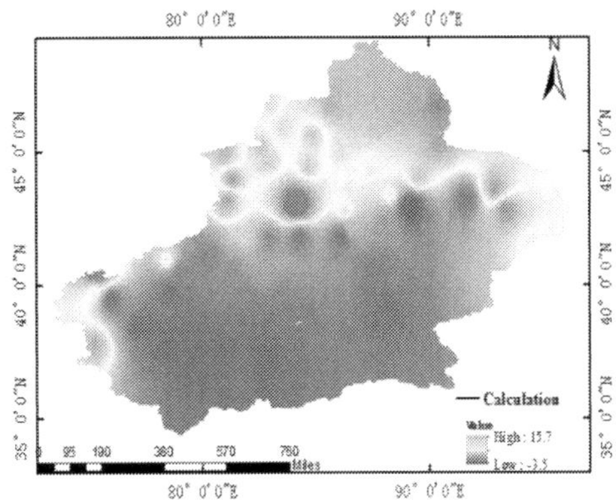


**Figure 5.** Spatial Distribution of Annual Average Temperature from 2002 to 2015

(1) Solar radiation and monsoon and other factors have influence on the temperature, precipitation and the growth of vegetation [5]. It is found that the correlation coefficients between annual maximum NDVI and annual precipitation and annual average temperature of 52 stations in the study areas are 0.634 ($P < 0.05$) and 0.279 (table 1) respectively, which indicates that the annual maximum NDVIs of vegetation have a higher positive correlation with annual precipitation and annual average temperature.

(2) Figure 6 is the scatter diagram between the annual maximum NDVI and annual precipitation. The annual average NDVI is of a significant positive correlation with the annual average precipitation. When the annual precipitation ranges from 100 to 150 mm, the distribution of the point is denser, which indicates that in this period of time, the vegetation is of good growth conditions with high vegetation coverage. This is the sensitive range of annual Maximum NDVI to annual average precipitation (figure 6).

(3)Figure 7 is the scatter diagram between the annual maximum NDVI and annual average temperature. The annual average NDVI is of a positive correlation with the annual average temperature. With the increase of the annual average temperature, the annual average NDVI increases, therefore the vegetation coverage increases gradually. When the annual average temperature is 7°C, the vegetation is of the best growth conditions; when the annual average temperature ranges from 6 to 7.5°C, the vegetation is exuberant. This is the sensitive range of annual Maximum NDVI to annual average temperature (figure7).

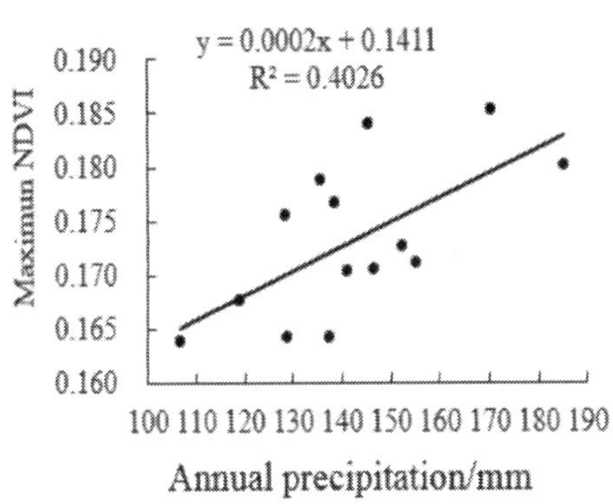


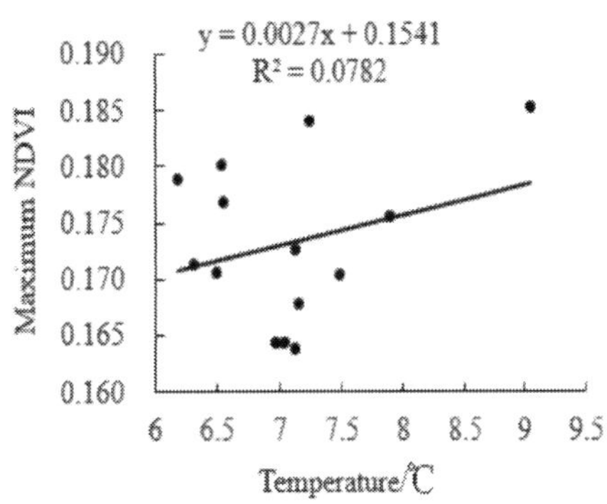

**Figure 6.** Correlation Analysis of Annual MaximumNDVI and Annual Precipitation

**Figure 7.** Correlation Analysis of Annual Maximum NDVI and Annual Average Temperature

**Table 1.** Correlation of Average Annual Maximum NDVI with Temperature and Precipitation of Xinjiang from 2002 to 2015

| NDVI | Climatic Factors | Correlation Coefficients | P |
|---|---|---|---|
| Annual maximum NDVI | Precipitation | 0.634* | 0.015 |
| | Temperature | 0.279 | 0.334 |

Note: "*" means significant correlation

## 5. Conclusion

(1) The annual maximum NDVI of Xinjiang has a low sensitivity to annual average temperature, while it has a high sensitivity to precipitation. In general, it has a significant correlation with climate, and extreme climatic events have a significant influence on the NDVI in the arid region of Xinjiang.
(2) Climate change is an important factor influencing the temporal and spatial variations of vegetation in the arid region of Xinjiang. In this study, only two climatic factors, temperature and precipitation, are studied, while the influence of other climatic factors including illumination intensity and sunshine duration on the vegetation coverage will be the emphasis in next study.
(3) Climatic factors as well as other driving factors, including urbanized life and human activities that increased rapidly as well as society and economy factors, will influence the vegetation. In human life, the development of oasis agriculture in Xinjiang, the change of plantation structure and the cultivated area of main crop (cotton) will influence the change of vegetation cover in Xinjiang to some degree. The guidance and analysis on the oasis agriculture and the ecological environment construction in Xinjiang will be more comprehensive if we try to use other geo-statistic methods to explore and quantify the interactive intensities of these factors, and carry out study for a longer time series.

## References

[1] Zhang Q, Yuan X L, Chen X, Luo G P and Li L H 2016 Vegetation Change and Its Response to Climate Change in Central Asia from 1982 to 2012 *Chinese Journal of Plant Ecology* **40** 13-23 (In Chinese with English Abstract)
[2] Gao Z Q and Liu H Y 2000 The Study on Driving Factors and Models of NDVI Change Based on Remote Sensing and GIS in China *Climatic and Environmental Research* **5** 155-64 (In Chinese with English Abstract)
[3] Yan X X and Dai C X 2013 Change Characteristics of the Vegetation Index in the North of the Tianshan Mountains and the Relation with the Temperature and Precipitation *Meteorological and Environmental Sciences* **36** 42-46 (In Chinese with English Abstract)
[4] Li J, Liu H B, Li L and Li C Y 2016 Vegetation Dynamic Changes in Northwest China based on Multi-source Remote Sensing Datasets in Recent 30 years *Arid Land Geography* **39** 387-94 (In Chinese with English Abstract)
[5] Dai S P, Zhang B and Wang H J 2010 Spatio-temporal Change of Vegetation Index NDVI in Northwest China and Its Influencing Factors *Journal of Geo-Information* **12** 315-21 (In Chinese with English Abstract)
[6] Liu X F, Zhou X F, Pan Y Z, Li Y Z and Zhao A Z 2015 Spatiotemporal Changes in Vegetation Coverage in China During 1982-2012 *Acta Ecologica Sinica* **35** 5331-42 (In Chinese with English Abstract)
[7] Shen X J, Zhou D W, Li F and Zhang H Y 2015 Vegetation Change and Its Response to Climate Change in Grassland Region of China *Scientia Geographica Sinica* **35** 622-29 (In Chinese with English Abstract)

# Glacier retreat of the Tian Shan and its impact on the urban growth and environment evaluated from satellite remote sensing data

**B H Fu[1*], Q Guo[1, 2], F Yan[1], J Zhang[1], P L Shi[1], M Ayinuer[1] and G L Xue[1]**

[1]Institute of Remote Sensing and Digital Earth, Chinese Academy of Science, Beijing 100094, China;
[2]University of Chinese Academy of Sciences, Beijing 100094, China

E-mail: fubh@radi.ac.cn

**Abstract.** The retreat of mountain glaciers, notably in high Asia, provides evidence for the rise of global temperature. The mass balance is vital for the health of a glacier. If the amount of frozen precipitation in the accumulation zone exceeds the quantity of glacial ice loss due to melting or lies in the ablation zone, the glacier will advance. Conversely, if the accumulation is less than the ablation, the glacier will retreat. Glaciers in retreat will have negative mass balances, and if they do not reach an equilibrium between accumulation and ablation, will eventually disappear. Long-term changes of the mountain glaciers in the Tian Shan, Central Asia, are not well constrained. Analyses of satellite remote sensing data combined with the ground observations reveal a 37.5% decline of glaciered area from 1989 to 2014 in No.1 Glacier, the headwaters of the Urumqi River basin, Chinese Tian Shan, which could be linked to increased summer melting. The results show that the area of glaciers was reduced from 31.55 $km^2$ in 18 August 1989 to 28.66 $km^2$ in 24 August 1994 and 19.74 $km^2$ in 31 August 2014. The glacier area was reduced by 0.47 $km^2$/per year in recent 25 years since 1989, and the annual reduction was 1.5%. Meanwhile, the urban area of Urumqi, the biggest city of Xinjiang Uygur Autonomous Region, increased from 156 $km^2$ in 1989 to 555 km2 in 2014. Correspondingly, the population of permanent residents increased from 1.06 million in 1989 to 3.53 million in 2014. We suggest that the decline of glacier area is driven primarily by summer melting and, possibly, linked to the combined effects of the global rise in temperatures and black carbon/$CO_2$ emission from coal-fired power plants, cement plants and petroleum chemical plants from the nearby Urumqi and surrounding regions. The continuing retreat of glaciers will have a number of different quantitative impacts. Populations in the arid Central Asia regions are heavily dependent on snow and glacier melt for their irrigation and drinking supplies. A continuation of the current retreat will eventually deplete the glacial ice and substantially reduce or eliminate runoff. It will also affect the availability of fresh water for mountain recreation, animals and plants that depend on glacier-melting.

## 1. Introduction

Central Asia is one of the world's most vulnerable regions to climate change. The arid climate, combined with a history of environmental mismanagement and limited climate-related risk mitigation strategies,

* Corresponding author.

have created a region that is increasingly vulnerable to the effects of rising temperatures, changing rainfall patterns, and increasing of aridity and frequency of extreme events. Since glaciers are highly sensitive to temperature increases, their changes are also considered as key indicators of some consequences of global warming.

Glaciers play an important role in the water cycle of Central Asia. Snow and glacier melting from the Tian Shan is essential for the water supply in the arid areas [1], such as Kazakhstan, Kyrgyzstan, Uzbekistan, and the Xinjiang Uyghur Autonomous Region, northwest China. Despite this importance, only little was known about how glaciers in this region changed in the past decades. In the last 50 years, the glaciers of Central Asia are estimated to have shrunk by between 20 to 30%. Glaciers of Central Asia are disappearing at rather high rates [2].

Most of the direct monitoring projects on the glaciers were shut down with the collapse of the Soviet Union except for the limited ground observations in China, though these modern observations often cover a limited region [2,3].

In the context of remote sensing, change detection refers to the process of identifying differences in the state of surface features by observing them at different times. The goal of change detection is generally an image that highlights areas that have changed between two (or more) time periods and the direction and magnitude of change.

Retreat of glaciers is likely to aggravate existing problems and pose additional risks to sustainable development, energy and food security in the arid area like Central Asia. In this study, we document the change of glaciers in upper reach of Urumqi River basin, Tian Shan and urban sprawl of Urumqi City, capital city of Xinjiang Uygur Autonomous Region over the past decades using the multi-temporal remote sensing. Emphasis is placed on results from the optical remote sensing interpretations and ground observations.

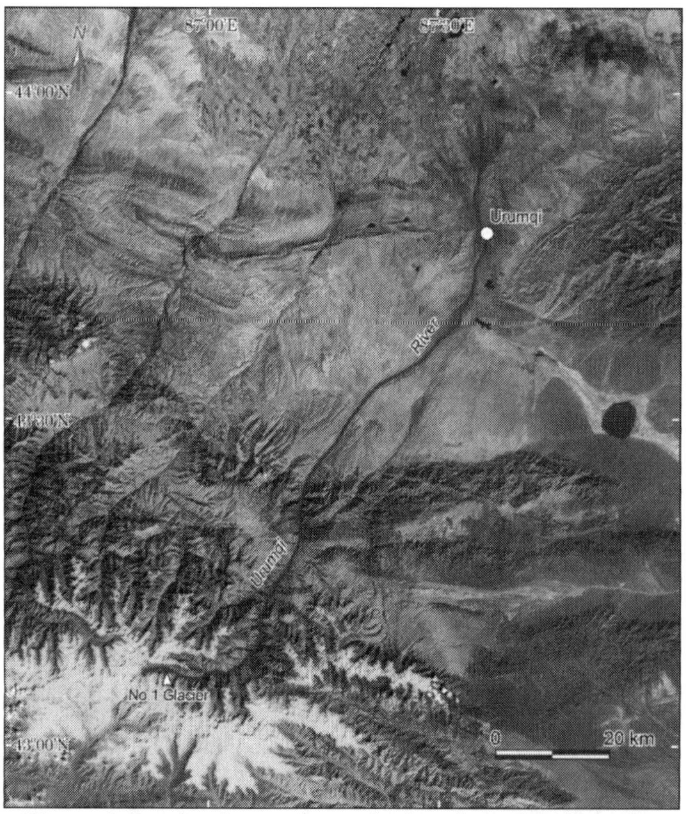


**Figure 1.** Landsat image shows the location of study region

## 2. Study area and methods

The No.1 glacier, one of the largest ice bodies in the Chinese Tian Shan, is located in in the headwaters of the Urumqi River, Tian Shan, China (Figure 1). Being the main source of the Urumqi River, it attracts thousands of visitors every year. The No.1 glacier is a vital source of freshwater storage and water supply, especially during the summer season for a large human population living downstream. There are around 15 glaciers exist in the headwater of the Urumqi River covering approximately 38 $km^2$ [4]. The 214 km-long Urumqi River system with its tributaries covers an area of 4684 $km^2$ [5].

*2.1. Detecting glacier changes and urban sprawl from Landsat time-series data during 1989 to 2014*
In this study, cloud-free full scenes of path 143, row 30, and path 143, row 29 acquired by Landsat series were used (see Table1).

**Table 1.** List of Landsat time-series data used in this study

| Satellite Data | Path/Row | Date of acquisition | Spatial resolution (m) |
|---|---|---|---|
| Landsat TM | 143/30 | 18 August 1989 | 30 |
| Landsat ETM+ | 143/30 | 24 August 1994 | 30 |
| Landsat 8 OLI | 143/30 | 31 August 2014 | 30 |
| Landsat TM | 143/29 | 18 September 1989 | 30 |
| Landsat 8 OLI | 143/29 | 31 August 2014 | 30 |

As shown in Figure 2, retreat of glaciers in the source region of Urumqi River is notably identifiable in the Landsat time-series images with around 30m resolution. We can calculate the area of glaciers in the upstream region of Urumqi River basin with an area of 1243 km2 using GIS software. The results show that the areas of glaciers are 31.55 $km^2$ in 18 August 1989, 28.66 $km^2$ in 24 August 1994 and 19.74 $km^2$ in 31 August 2014 (Figure 3), respectively. The analysis of early air-photos indicated that the area of glaciers is about 38 $km^2$ in the 1950s [4].

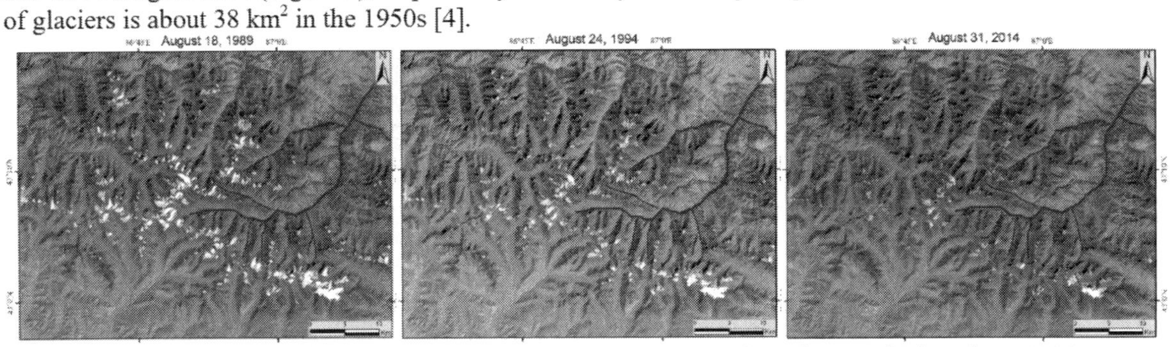


**Figure 2.** Landsat images show the glacier changes in the source region of Urumqi River from 1989 to 2014

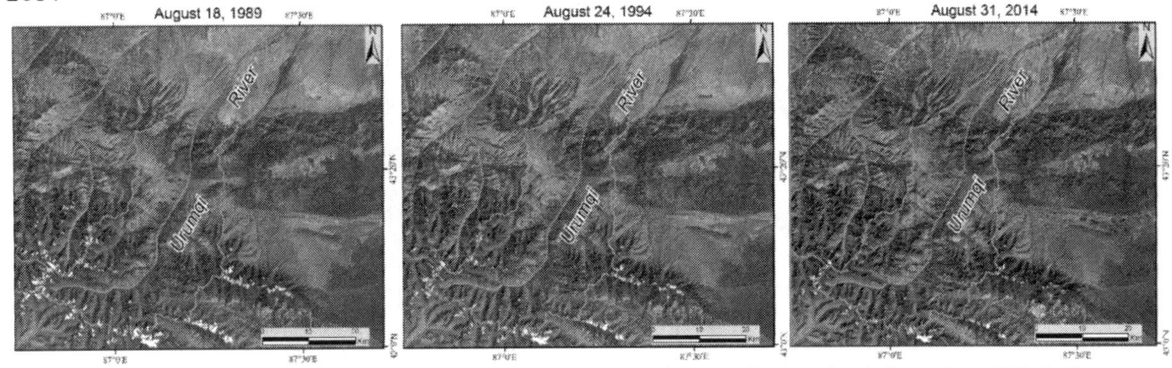


**Figure 3.** The changes of glaciered area in upstream region of Urumqi River basin based on GIS during 1989 to 2014

IOP Conf. Series: Earth and Environmental Science **74** (2017) 012022 doi:10.1088/1755-1315/74/1/012022

Meanwhile, we also investigate the urban sprawl of Urumqi city during 1989 to 2014. As we see from Figure 4, the urban land use grows rapidly in the same time span. The urban area expands from 156 $km^2$ in 1989 to 555 $km^2$ in 2014. At the same time, the population of permanent residents increased from 1.06 million in 1989 to 3.53 million in 2014 according to the Urumqi city yearbook.

### *2.2. Ground observations*

In early October 2014, we conducted a field investigation and measurements along the Tian Shan No.1 Glacier and Urumqi River, based on the area of interests detected with optical remote sensing as described above.

We found that the Tian Shan No.1 Glacier was experiencing a rapid retreating caused by summer climate warming in recent years. Our measurements showed that the retreating distance is about 5-6 meters during May 18 to October 6, 2014 (see Figure 5a). We are surprised that the glaciers are still melting in early October (see Figure 5b.)

In the past 30 years, the small glaciers of the Tian Shan have been disappeared due to continued glacier shrinkage (see Figure 6a).

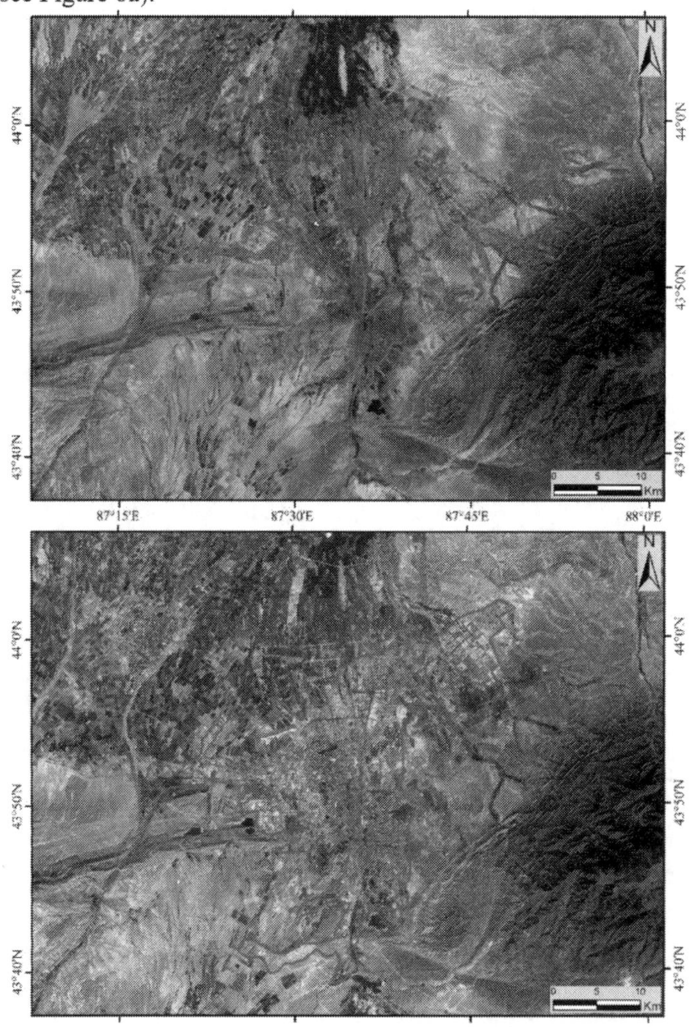


**Figure 4.** Landsat images of the Urumqi city show the urban sprawl from 1989 to 2014

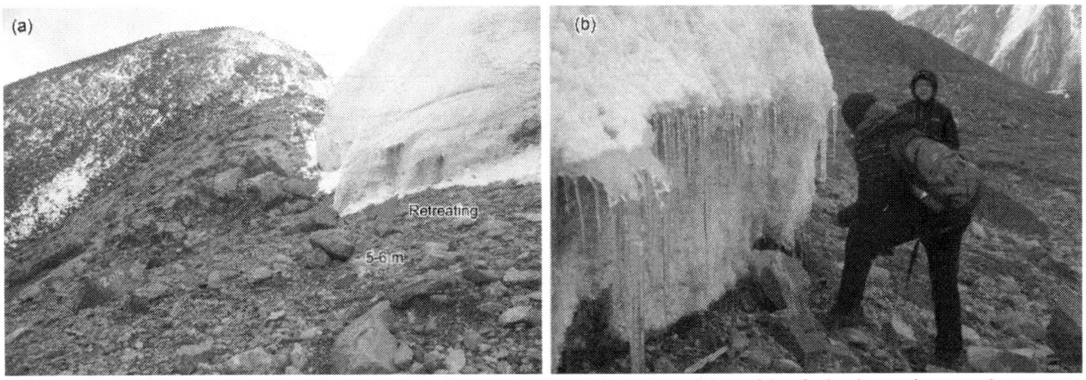


**Figure 5.** Field photographs show the rapid retreating (a) and melting (b) of glaciers observed at western branch of Tian Shan No.1 Glacier.

**Figure 6.** Field photographs show the bare rocks caused by the rapid retreating of glaciers (a) in the source region of Urumqi River and (b) $CO_2$, black carbon and thermal emissions from coal-fired power plant of glaciers observed at 30 km north to Tian Shan No.1 Glacier.

## 3. Discussion

Climate-driven changes in glacier-fed streamflow regimes have direct implications on freshwater supply, irrigation and hydropower potential. Reliable information about current and future glaciation and runoff is crucial for water allocation, a complex task in Central Asia, where the collapse of the Soviet Union has transformed previously interdependent republics into autonomous upstream and downstream countries. Although the impacts of climate change on glaciation and runoff have been addressed in previous work undertaken in the Tian Shan [3,5,6], a coherent, regional perspective of these findings has not been presented until now. Here we show that glacier shrinkage is most pronounced in peripheral, lower-elevation ranges near the densely populated forelands, where summers are dry and where snow and glacial meltwater is essential for water availability. Shifts of seasonal runoff maxima have already been observed in the Urumqi River. Based on the observations of runoff and precipitation in the Urumqi River basin, the variation features in climatic shift from warm dry to warm humid are analysed [5]. It is reveals that there is an obvious increasing trend of runoff at the Glacier No.1 and Yingxunqiao Hydrologic Stations of the Urumqi River after 1987. Especially, the runoff at the Glacier No.1 Hydrologic Stations has a very clear shift in 1997. Moreover, analyses of long-term climate and glacier records over the past 45 years (1959 to 2003) in Urumqi River source region, Tian Shan show that summer temperature and annual precipitation near the glacier increased by 0.8°C and 87 mm (19%), respectively. The glacier continuously retreated during the same period, with the cumulated mass balance being 10,032 mm, or 20% of the glacier volume. Annual basin runoff has significantly increased by 413 mm or 62% during 1980–2003 due to precipitation increase and enhanced glacier melt caused by summer climate warming. Both summer precipitation and temperate are negatively correlated with mass balance and positively associated with runoff. Relative to precipitation mass balance relation, the regression between temperature and mass balance is much stronger, indicating that summer temperature controls glacier mass balance and runoff changes [3].

The main reason of the increasing runoff in the hydrologic stations of upstream region is due to the two factors: strong melting of glacier and durative increase in precipitation in the source area of the Urumqi River [3,5]. Moreover, many recent studies also revealed that industrial black carbon has caused the melting of glaciers [7,8] as we observed in the field (Figure 5b and 6b).

The increasing water consumption of a growing population is likely to put additional pressure on both food and energy supplies, and could aggravate tensions within the region and between Central Asia and neighbouring countries. Currently observed changes in glaciers threaten the region in the short term with floods and with reduced water resources availability in the long term. However, it is suggested that summer runoff will further decrease in these rivers if precipitation and discharge from thawing permafrost bodies do not compensate sufficiently for water shortfalls.

Effective or accurate monitoring of the glaciers rate of degradation and their current status are essential to minimizing the negative effects of this degradation through better planning and management of water resources.

## 4. Conclusions

The situation is of particular concern in light of both the local population growth and the continued glacier shrinkage anticipated in response to climatic changes. Our study demonstrated that multi-temporal Landsat series optical remote sensing data were particularly useful for capturing surface change at sufficient frequency to study glacier dynamics and changes for the urban region.

We combined geospatial analysis techniques with remote-sensing and field data to document spatial patterns of glacier changes in the headwaters of Urumqi River basin in the past 25 years (from 1989 to 2014).

We summarize as follows:

1. Glaciers of the source region of Urumqi River lost 37.5 % (11.81 $km^2$) of their area from 1989 to 2014. Glaciered area was reduced from 31.55 $km^2$ in 18 August 1989 to 28.66 $km^2$ in 24 August 1994 and 19.74 $km^2$ in 31 August 2014. The glacier area was reduced by 0.47 $km^2$/per year, and the annual reduction was 1.5%.

2. The urban area of Urumqi, the capital city of Xinjiang Uygur Autonomous Region, increased from 156 $km^2$ in 1989 to 555 $km^2$ in 2014. Correspondingly, the population of permanent residents increased from 1.06 million in 1989 to 3.53 million in 2014.

3. We suggest that the decline was driven primarily by summer melting and, possibly, linked to the combined effects of the global rise in temperatures and carbon/$CO_2$ emission from coal-fired power plants, cement plants and petroleum chemical plants from the nearby Urumqi and surrounding regions.

## References

[1] Yao T D, Wang Y Q, Liu S Y, Pu J C, Shen, Y P, and Lu A 2004 Recent glacial retreat in High Asia in China and its impact on water resource in northwest *China Science in China (Series D Earth Sciences)* **47** 1065-75

[2] Farinotti D, Longuevergne L, Moholdt G, Duethmann D, Mölg T, Bolch T, Vorogushyn S, Güntner A 2015 Substantial glacier mass loss in the Tien Shan over the past 50 years *Nature Geoscience* **8** 716-22

[3] Ye B S, Yang D Q, Jiao K Q, Han T D, Jin Z F, Yang H A, and Li Z Q 2005 The Urumqi River source Glacier No. 1, Tianshan, China: Changes over the past 45 years *Geophysical Research Letters* **32** 154-64

[4] Zhang W C, Chen J, Ogawa K, and Yamaguchi Y, 2005 An approach to estimating evapotranspiration in the Urumqi River basin, Tianshan, China, by means of remote sensing and a geographical information system technique *Hydrological Processes* **19** 1839-54

[5] Han T D, Ye B S, Ding Y J, Jiao K Q 2005 Increasing runoff in the Urumqi river basin since

1980s *Journal of Glaciology and Geocryology* **27** 655-59 (in Chinese)
[6] Diao P, Lu B, Yang X L 2000 Features of climate and glacier change in source of Urumqi River and its effect on runoff *Xinjiang Meteorology* **23** 10-12
[7] Menon S, Koch D, Beig G, Sahu S, Fasullo J., and Orlikowski D 2009 Black carbon aerosols and the third polar ice cap *Atmos. Chem. Phys. Discuss.* **9** 26593-625
[8] Xu B Q, Cao J J, Hansen J, Yao T D, Joswia D R, Wang N L, Wu G J, Wang M, Zhao H B, Yang W, Liu X Q, and Jianqiao He J Q 2009 Black soot and the survival of Tibetan glaciers *PNAS* **106** 22114-18